Genetics: A Conceptual Approach

Genetics: A Conceptual Approach

Malcolm Kripke

SYRAWOOD
PUBLISHING HOUSE
New York

Published by Syrawood Publishing House,
750 Third Avenue, 9th Floor,
New York, NY 10017, USA
www.syrawoodpublishinghouse.com

Genetics: A Conceptual Approach
Malcolm Kripke

International Standard Book Number: 978-1-64740-012-5 (Hardback)

Cataloging-in-Publication Data

Genetics : a conceptual approach / Malcolm Kripke.
p. cm.
Includes bibliographical references and index.
ISBN 978-1-64740-012-5
1. Genetics. 2. Biology. 3. Embryology. 4. Genomes. I. Kripke, Malcolm.
QH430 .G46 2020
576.5--dc23

Printed in China.

TABLE OF CONTENTS

PREFACE

The branch of biology which deals with the study of genetic variation, genes and heredity in organisms is known as genetics. The primary principles of genetics are trait inheritance and molecular inheritance mechanisms of genes. It is also involved in the study of the function and behavior of genes. Genetics studies the gene structure, function, variation and distribution in various contexts such as the cell, organism and population. Some of the major sub-fields within this discipline are epigenetics and population genetics. It examines the genetic processes that work with an organism's environment and experiences in order to influence its development and behavior. This textbook presents the complex subject of genetics in the most comprehensible and easy to understand language. It is a valuable compilation of topics, ranging from the basic to the most complex theories and principles in this field. For all those who are interested in genetics, this book can prove to be an essential guide.

A short introduction to every chapter is written below to provide an overview of the content of the book:

Chapter 1 - Genetics is a branch of biology that deals with the study of genetic variation, heredity and genes in organisms. Ecogenetics, developmental genetics, microbial genetics are the various branches that fall under this domain. It also deals with the study of genetic materials like nucleic acid and genome. This is an introductory chapter which will introduce briefly all these significant aspects of genetics.; **Chapter 2** - The passing on of traits from parents to their offspring is known as heredity. It takes place either through sexual reproduction or asexual reproduction. A chromosome is a DNA molecule that consists of a part or all the genetic material of an organism. This chapter discusses in detail the concepts and processes related to heredity and chromosomes such as the behavior of chromosomes during cell division, chromosome mutations and Mendelian inheritance.; **Chapter 3** - DNA is a molecule which is made up of two chains that coil around each other in a double helix formation. RNA is also a chain of nucleotides but exists as a single strand. The topics elaborated in this chapter will help in gaining a better perspective about the different structures, functions and processes associated with DNA and RNA.; **Chapter 4** - The process by which information from a gene is used in the synthesis of a functional gene product is known as gene expression. The various steps which are a part of gene expression are transcription, translation and post-translational modification of protein. These different components of gene expression as well as gene mutation have been thoroughly discussed in this chapter.; **Chapter 5** - Gene regulation is a process that includes a wide range of mechanisms which helps cells to increase or decrease the production of specific gene products. It can be applied to different steps of gene expression such as transcription, translation and post translation. The chapter closely examines these key aspects of gene regulation to provide an extensive understanding of the subject.; **Chapter 6** - The direct manipulation of an organism's genes using biotechnology is known as genetic engineering. Different organisms are modified using genetic engineering such as bacteria and virus. Genetic engineering is also involved in genetically modifying crops and providing gene therapy. All these diverse applications of gene therapy have been carefully analyzed in this chapter.

Finally, I would like to thank my fellow scholars who gave constructive feedback and my family members who supported me at every step.

Malcolm Kripke

Chapter 1

Genetics: An Introduction

Genetics is a branch of biology that deals with the study of genetic variation, heredity and genes in organisms. Ecogenetics, developmental genetics, microbial genetics are the various branches that fall under this domain. It also deals with the study of genetic materials like nucleic acid and genome. This is an introductory chapter which will introduce briefly all these significant aspects of genetics.

Genetics is the study of heredity in general and of genes in particular. Genetics forms one of the central pillars of biology and overlaps with many other areas, such as agriculture, medicine, and biotechnology.

Since the dawn of civilization, humankind has recognized the influence of heredity and applied its principles to the improvement of cultivated crops and domestic animals. A Babylonian tablet more than 6,000 years old, for example, shows pedigrees of horses and indicates possible inherited characteristics. Other old carvings show cross-pollination of date palm trees. Most of the mechanisms of heredity, however, remained a mystery until the 19th century, when genetics as a systematic science began.

Genetics arose out of the identification of genes, the fundamental units responsible for heredity. Genetics may be defined as the study of genes at all levels, including the ways in which they act in the cell and the ways in which they are transmitted from parents to offspring. Modern genetics focuses on the chemical substance that genes are made of, called deoxyribonucleic acid, or DNA, and the ways in which it affects the chemical reactions that constitute the living processes within the cell. Gene action depends on interaction with the environment. Green plants, for example, have genes containing the information necessary to synthesize the photosynthetic pigment chlorophyll that gives them their green colour. Chlorophyll is synthesized in an environment containing light because the gene for chlorophyll is expressed only when it interacts with light. If a plant is placed in a dark environment, chlorophyll synthesis stops because the gene is no longer expressed.

Genetics as a scientific discipline stemmed from the work of Gregor Mendel in the middle of the 19th century. Mendel suspected that traits were inherited as discrete units, and, although he knew nothing of the physical or chemical nature of genes at the time, his units became the basis for the development of the present understanding of heredity. All present research in genetics can be traced back to Mendel's discovery of the laws governing the inheritance of traits.

Areas of Study

Classical Genetics

Classical genetics, which remains the foundation for all other areas in genetics, is concerned primarily with the method by which genetic traits—classified as dominant (always expressed), recessive (subordinate to a dominant trait), intermediate (partially expressed), or polygenic (due to multiple genes)—are transmitted in plants and animals. These traits may be sex-linked (resulting from the action of a gene on the sex, or X, chromosome) or autosomal (resulting from the action of a gene on a chromosome other than a sex chromosome). Classical genetics began with Mendel's study of inheritance in garden peas and continues with studies of inheritance in many different plants and animals. Today a prime reason for performing classical genetics is for gene discovery—the finding and assembling of a set of genes that affects a biological property of interest.

Cytogenetics

Cytogenetics, the microscopic study of chromosomes, blends the skills of cytologists, who study the structure and activities of cells, with those of geneticists, who study genes. Cytologists discovered chromosomes and the way in which they duplicate and separate during cell division at about the same time that geneticists began to understand the behaviour of genes at the cellular level. The close correlation between the two disciplines led to their combination.

Plant cytogenetics early became an important subdivision of cytogenetics because, as a general rule, plant chromosomes are larger than those of animals. Animal cytogenetics became important after the development of the so-called squash technique, in which entire cells are pressed flat on a piece of glass and observed through a microscope; the human chromosomes were numbered using this technique.

Today there are multiple ways to attach molecular labels to specific genes and chromosomes, as well as to specific RNAs and proteins, that make these molecules easily discernible from other components of cells, thereby greatly facilitating cytogenetics research.

Microbial Genetics

Microorganisms were generally ignored by the early geneticists because they are small in size and were thought to lack variable traits and the sexual reproduction necessary for a mixing of genes from different organisms. After it was discovered that microorganisms have many different physical and physiological characteristics that are amenable to study, they became objects of great interest to geneticists because of their small size and the fact that they reproduce much more rapidly than larger organisms. Bacteria became important model organisms in genetic analysis, and many discoveries of general interest in genetics arose from their study. Bacterial genetics is the centre of cloning technology.

Viral genetics is another key part of microbial genetics. The genetics of viruses that attack bacteria were the first to be elucidated. Since then, studies and findings of viral genetics have been applied to viruses pathogenic on plants and animals, including humans. Viruses are also used as vectors (agents that carry and introduce modified genetic material into an organism) in DNA technology.

Molecular Genetics

Molecular genetics is the study of the molecular structure of DNA, its cellular activities (including its replication), and its influence in determining the overall makeup of an organism. Molecular genetics relies heavily on genetic engineering (recombinant DNA technology), which can be used to modify organisms by adding foreign DNA, thereby forming transgenic organisms. Since the early 1980s, these techniques have been used extensively in basic biological research and are also fundamental to the biotechnology industry, which is devoted to the manufacture of agricultural and medical products. Transgenesis forms the basis of gene therapy, the attempt to cure genetic disease by addition of normally functioning genes from exogenous sources.

Genomics

The development of the technology to sequence the DNA of whole genomes on a routine basis has given rise to the discipline of genomics, which dominates genetics research today. Genomics is the study of the structure, function, and evolutionary comparison of whole genomes. Genomics has made it possible to study gene function at a broader level, revealing sets of genes that interact to impinge on some biological property of interest to the researcher. Bioinformatics is the computer-based discipline that deals with the analysis of such large sets of biological information, especially as it applies to genomic information.

Population Genetics

The study of genes in populations of animals, plants, and microbes provides information on past migrations, evolutionary relationships and extents of mixing among different varieties and species, and methods of adaptation to the environment. Statistical methods are used to analyze gene distributions and chromosomal variations in populations.

Population genetics is based on the mathematics of the frequencies of alleles and of genetic types in populations. For example, the Hardy-Weinberg formula, $p^2 + 2pq + q^2 = 1$, predicts the frequency of individuals with the respective homozygous dominant (AA), heterozygous (Aa), and homozygous recessive (aa) genotypes in a randomly mating population. Selection, mutation, and random changes can be incorporated into such mathematical models to explain and predict the course of evolutionary change at the population level. These methods can be used on alleles of known phenotypic effect, such as the recessive allele for albinism, or on DNA segments of any type of known or unknown function.

Human population geneticists have traced the origins and migration and invasion routes of modern humans, Homo sapiens. DNA comparisons between the present peoples on the planet have pointed to an African origin of Homo sapiens. Tracing specific forms of genes has allowed geneticists to deduce probable migration routes out of Africa to the areas colonized today. Similar studies show to what degree present populations have been mixed by recent patterns of travel.

Behaviour Genetics

Another aspect of genetics is the study of the influence of heredity on behaviour. Many aspects of animal behaviour are genetically determined and can therefore be treated as similar to other biological properties. This is the subject material of behaviour genetics, whose goal is to determine which genes control various aspects of behaviour in animals. Human behaviour is difficult to analyze because of the powerful effects of environmental factors, such as culture. Few cases of genetic determination of complex human behaviour are known. Genomics studies provide a useful way to explore the genetic factors involved in complex human traits such as behaviour.

Human Genetics

Some geneticists specialize in the hereditary processes of human genetics. Most of the emphasis is on understanding and treating genetic disease and genetically influenced ill health, areas collectively known as medical genetics. One broad area of activity is laboratory research dealing with the mechanisms of human gene function and malfunction and investigating pharmaceutical and other types of treatments. Since there is a high degree of evolutionary conservation between organisms, research on model organisms—such as bacteria, fungi, and fruit flies (Drosophila)—which are easier to study, often provides important insights into human gene function.

Many single-gene diseases, caused by mutant alleles of a single gene, have been discovered. Two well-characterized single-gene diseases include phenylketonuria (PKU) and Tay-Sachs disease. Other diseases, such as heart disease, schizophrenia, and depression, are thought to have more complex heredity components that involve a number of different genes. These diseases are the focus of a great deal of research that is being carried out today.

Another broad area of activity is clinical genetics, which centres on advising parents of the likelihood of their children being affected by genetic disease caused by mutant genes and abnormal chromosomestructure and number. Such genetic counseling is based on examining individual and family medical records and on diagnostic procedures that can detect unexpressed, abnormal forms of genes. Counseling is carried out by physicians with a particular interest in this area or by specially trained nonphysicians.

Methods in Genetics

Experimental Breeding

Genetically diverse lines of organisms can be crossed in such a way to produce different combinations of alleles in one line. For example, parental lines are crossed, producing an F_1 generation, which is then allowed to undergo random mating to produce offspring that have purebreeding genotypes (i.e., AA, bb, cc, or DD). This type of experimental breeding is the origin of new plant and animal lines, which are an important part of making laboratory stocks for basic research. When applied to commerce, transgenic commercial lines produced experimentally are called genetically modified organisms (GMOs). Many of the plants and animals used by humans today (e.g., cows, pigs, chickens, sheep, wheat, corn (maize), potatoes, and rice) have been bred in this way.

Cytogenetic Techniques

Cytogenetics focuses on the microscopic examination of genetic components of the cell, including chromosomes, genes, and gene products. Older cytogenetic techniques involve placing cells in paraffin wax, slicing thin sections, and preparing them for microscopic study. The newer and faster squash technique involves squashing entire cells and studying their contents. Dyes that selectively stain various parts of the cell are used; the genes, for example, may be located by selectively staining the DNA of which they are composed. Radioactive and fluorescent tags are valuable in determining the location of various genes and gene products in the cell. Tissue-culture techniques may be used to grow cells before squashing; white blood cells can be grown from samples of human blood and studied with the squash technique. One major application of cytogenetics in humans is in diagnosing abnormal chromosomal complements such as Down syndrome (caused by an extra copy of chromosome 21) and Klinefelter syndrome (occurring in males with an extra X chromosome). Some diagnosis is prenatal, performed on cell samples from amniotic fluid or the placenta.

Biochemical Techniques

Biochemistry is carried out at the cellular or subcellular level, generally on cell extracts. Biochemical methods are applied to the main chemical compounds of genetics—notably DNA, RNA, and protein. Biochemical techniques are used to determine the activities of genes within cells and to analyze substrates and products of gene-controlled reactions. In one approach, cells are ground up and the substituent chemicals are fractionated for further analysis. Special techniques (e.g., chromatography and electrophoresis) are used to separate the components of proteins so that inherited differences in their structures can be revealed. For example, more than 100 different kinds of human hemoglobin molecules have been identified. Radioactively tagged compounds are valuable in studying the biochemistry of whole cells. For example, thymine is a compound found only in DNA; if radioactive thymine is placed in a tissue-culture medium in which cells are growing, genes use it to duplicate themselves. When cells containing radioactive thymine are analyzed, the results show that, during duplication, the DNA molecule splits in half, and each half synthesizes its missing components.

Chemical tests are used to distinguish certain inherited conditions of humans; e.g., urinalysis and blood analysis reveal the presence of certain inherited abnormalities—phenylketonuria (PKU), cystinuria, alkaptonuria, gout, and galactosemia. Genomics has provided a battery of diagnostic tests that can be carried out on an individual's DNA. Some of these tests can be applied to fetuses in utero.

Physiological Techniques

Physiological techniques, directed at exploring functional properties or organisms, are also used in genetic investigations. In microorganisms, most genetic variations involve some important cell function. Some strains of one bacterium (Escherichia coli), for example, are able to synthesize the vitamin thiamin from simple compounds; others, which lack an enzyme necessary for this synthesis, cannot survive unless thiamin is already present. The two strains can be distinguished by placing them on a thiamin-free mixture: those that grow have the gene for the enzyme, those that fail to grow do not. The technique also is applied to human cells, since many inherited human abnormalities are caused by a faulty gene that fails to produce a vital enzyme; albinism, which results from an inability to produce the pigment melanin in the skin, hair, or iris of the eyes, is an example of an enzyme deficiency in man.

Molecular Techniques

Although overlapping with biochemical techniques, molecular genetics techniques are deeply involved with the direct study of DNA. This field has been revolutionized by the invention of recombinant DNA technology. The DNA of any gene of interest from a donor organism (such as a human) can be cut out of a chromosome and inserted into a vector to make recombinant DNA, which can then be amplified and manipulated, studied, or used to modify the genomes of other organisms by transgenesis. A fundamental step in recombinant DNA technology is amplification. This is carried out by inserting the recombinant DNA molecule into a bacterial cell, which replicates and produces many copies of the bacterial genome and the recombinant DNA molecule (constituting a DNA clone). A collection of large numbers of clones of recombinant donor DNA molecules is called a genomic library. Such libraries are the starting point for sequencing entire genomes such as the human genome. Today genomes can be scanned for small molecular variants called single nucleotide polymorphisms, or SNPs ("snips"), which act as chromosomal tags to associated specific regions of DNA that have a property of interest and may be involved in a human disease or disorder.

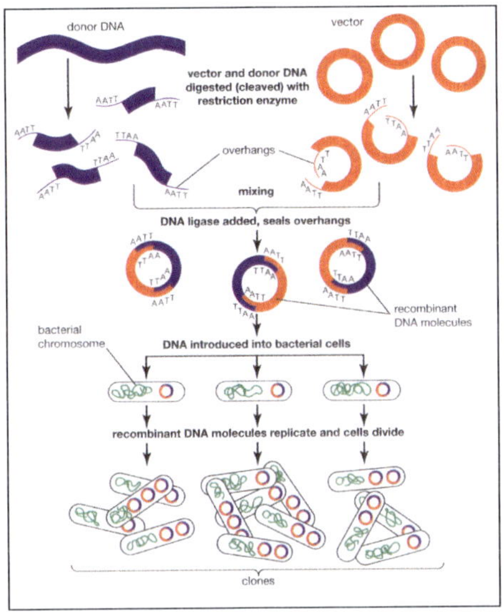

Steps involved in the engineering of a recombinant DNA molecule.

Immunological Techniques

Many substances (e.g., proteins) are antigenic; i.e., when introduced into a vertebrate body, they stimulate the production of specific proteins called antibodies. Various antigens exist in red blood cells, including those that make up the major blood groups of man (A, B, AB, O). These and other antigens are genetically determined; their study constitutes immunogenetics. Blood antigens of man include inherited variations, and the particular combination of antigens in an individual is almost as unique as fingerprints and has been used in such areas as paternity testing (although this approach has been largely supplanted by DNA-based techniques).

Immunological techniques are used in blood group determinations in blood transfusions, in organ transplants, and in determining Rhesus incompatibility in childbirth. Specific antigens of the human leukocyte antigen (HLA) genes are correlated with human diseases and disease predispositions. Antibodies also have a genetic basis, and their seemingly endless ability to match any antigen presented is based on special types of DNA shuffling processes between antibody genes. Immunology is also useful in identifying specific recombinant DNA clones that synthesize a specific protein of interest.

Mathematical Techniques

Because much of genetics is based on quantitative data, mathematical techniques are used extensively in genetics. The laws of probability are applicable to crossbreeding and are used to predict frequencies of specific genetic constitutions in offspring. Geneticists also use statistical methods to determine the significance of deviations from expected results in experimental analyses. In addition, population genetics is based largely on mathematical logic—for example, the Hardy-Weinberg equilibrium and its derivatives.

Bioinformatics uses computer-centred statistical techniques to handle and analyze the vast amounts of information accumulating from genome sequencing projects. The computer program scans the DNA looking for genes, determining their probable function based on other similar genes, and comparing different DNA molecules for evolutionary analysis. Bioinformatics has made possible the discipline of systems biology, treating and analyzing the genes and gene products of cells as a complete and integrated system.

Applied Genetics

Medicine

Genetic techniques are used in medicine to diagnose and treat inherited human disorders. Knowledge of a family history of conditions such as cancer or various disorders may indicate a hereditary tendency to develop these afflictions. Cells from embryonic tissues reveal certain genetic abnormalities, including enzyme deficiencies, that may be present in newborn babies, thus permitting early treatment. Many countries require a blood test of newborn babies to determine the presence of an enzyme necessary to convert an amino acid, phenylalanine, into simpler products. Phenylketonuria (PKU), which results from lack of the enzyme, causes permanent brain damage if not treated soon after birth. Many different types of human genetic diseases can be detected in embryos as young as 12 weeks; the procedure involves removal and testing of a small amount of fluid from around the embryo (called amniocentesis) or of tissue from the placenta (called chorionic villus sampling).

Gene therapy is based on modification of defective genotypes by adding functional genes made through recombinant DNA technology. Bioinformatics is being used to "mine" the human genome for gene products that might be candidates for designer pharmaceutical drugs.

Agriculture and Animal Husbandry

Agriculture and animal husbandry apply genetic techniques to improve plants and animals. Breeding analysis and transgenic modification using recombinant DNA techniques are routinely used. Animal breeders use artificial insemination to propagate the genes of prize bulls. Prize cows can transmit their genes to hundreds of offspring by hormone treatment, which stimulates the release of many eggs that are collected, fertilized, and transplanted to foster mothers. Several types of mammals can be cloned, meaning that multiple identical copies can be produced of certain desirable types.

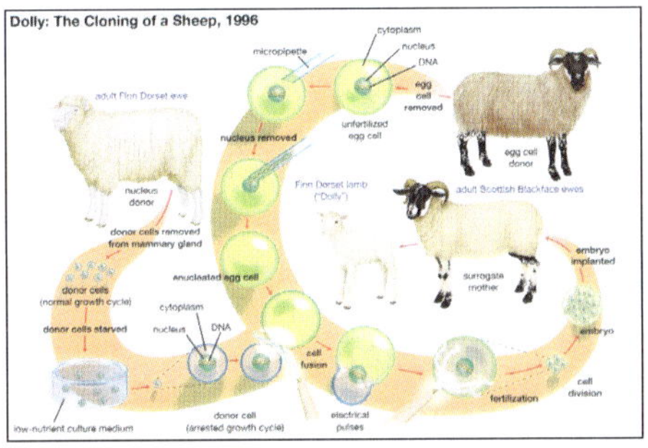

Dolly the sheep was successfully cloned in 1996 by fusing the nucleus from a mammary-gland cell of a Finn Dorset ewe into an enucleated egg cell taken from a Scottish Blackface ewe. Carried to term in the womb of another Scottish Blackface ewe, Dolly was a genetic copy of the Finn Dorset ewe.

Plant geneticists use special techniques to produce new species, such as hybrid grains (i.e., produced by crossing wheat and rye), and plants resistant to destruction by insect and fungal pests.

Plant breeders use the techniques of budding and grafting to maintain desirable gene combinations originally obtained from crossbreeding. Transgenic plant cells can be made into plants by growing the cells on special hormones. The use of the chemical compound colchicine, which causes chromosomes to double in number, has resulted in many new varieties of fruits, vegetables, and flowers. Many transgenic lines of crop plants are commercially advantageous and are being introduced into the market.

Industry

Various industries employ geneticists; the brewing industry, for example, may use geneticists to improve the strains of yeast that produce alcohol. The pharmaceutical industry has developed strains of molds, bacteria, and other microorganisms high in antibiotic yield. Penicillin and cyclosporin from fungi, and streptomycin and ampicillin from bacteria, are some examples.

Biotechnology, based on recombinant DNA technology, is now extensively used in industry. "Designer" lines of transgenic bacteria, animals, or plants capable of manufacturing some commercial product are made and used routinely. Such products include pharmaceutical drugs and industrial chemicals such as citric acid.

Ecogenetics

Ecogenetics is a branch of genetics and environmental science that studies how genetics affects responses to substances found within a given environment. It includes both population-level studies that look at how the general human genome is affected by or otherwise interacts with a given substance, as well as studies that investigate how variation within the human genome affects individual human responses to a given substance.

Ecogenetic methodologies provide occupational health researchers with another way to understand the manner in which exposure to a hazardous substance affects worker health. This allows for better risk assessments and a more accurate understanding of what threshold limits are necessary to protect worker health.

Ecogenetics provides insight into the risk level that substances pose to individuals by advancing our understanding of how genetics exacerbates or inhibits the negative effects of that substance. A chief interest of ecogeneticists involves genetic polymorphism, which refers to the genetic variation that exists between humans, and the impact it has on how humans interact with chemical substances.

A better understanding of the effects of genetic variation could conceivably protect genetically susceptible individuals from being harmed by exposure to an amount of substance that can be well-tolerated by the majority of the population. For instance, a 2015 Harvard study successfully discovered a genetic link between occupational noise exposure and hearing loss that could be used to provide specific guidelines for protecting susceptible individuals.

The use of genetic screening in a workplace context is a complex matter that coincides with issues related to worker privacy and genetic discrimination, and as such, genetic screening programs are difficult to implement. Genetic testing has been implemented by some companies in the past, but it is not currently a regular part of most OHS programs.

The use of genetic screening to inform workplace health and safety is not required by occupational health and safety authorities, and there are currently no validated genetic tests for assessing increased workplace risk to a given chemical. The National Institute of Occupational Safety and Health (NIOSH) anticipates that the use of genetic tests will become more prevalent in the future as ecogenetic science advances and as relevant testing becomes easier and more affordable.

Ecogenetics principally deals with effects of preexisting genetically-determined variability on the response to environmental agents. The word environmental is defined broadly to include the physical, chemical, biological, atmospheric, and climate agents. Ecogenetics, therefore, is an all-embracing term, and concepts such as pharmacogenetics are seen as subcomponents of ecogenetics. This work grew logically from the book entitled Pollutants and High Risk Groups, which presented an overview of the various host factors i.e. age, heredity, diet, preexisting diseases, and lifestyles which affect environmentally-induced disease.

It should be emphasized the primary intention of ecogenetics is to provide an objective and critical evaluation of the scientific literature pertaining to genetic factors and differential susceptibility to environmental agents, with particular emphasis on those agents typically considered pollutants. It is important to realize though that ones genetic makeup, while important, is but one of an array of host factors contributing to overall adaptive capacity of the individual. In many instances, it is possible for such factors to interact in ways that may enhance or offset the effect of each other.

- Red blood cell conditions: There is a broad group of genetic diseases that result in either producing or predisposing affected individuals to the development of hemolytic anemias. These diseases include abnormal hemoglobin, inability to manufacture one or the other of the peptide globin chains of the hemoglobin, and deficiencies of the Embden-Meyerhoff monophosphate.
- Liver metabolism: Individuals lacking the ability to detoxify and excrete PCB's may have a high risk of total liver failure in conjunction with certain ecological conditions.
- Cardiovascular diseases: The pathologic lesion of atherosclerosis is a plaque-like substance that thickens the innermost and middle of the three layers of the artery wall. The thickening of the intimal and medial layers results from the accumulation of the proliferating smooth muscle cells that are encompassed by interstitial substances such as collagen, elastin, glycosamninoglycans, and fibrin.
- Respiratory diseases: There are three genetically-based respiratory diseases that can directly correspond with ecological functions and induce disease. These include lung cancer and the upper and lower respiratory tract associated with a serum Ig A deficiency.

Microbial Genetics

Microbial genetics is a subject area within microbiology and genetic engineering. Microbial genetics studies microorganisms for different purposes. The microorganisms that are observed are bacteria, and archaea. Some fungi and protozoa are also subjects used to study in this field. The studies of microorganisms involve studies of genotype and expression system. Genotypes are the inherited compositions of an organism. Genetic Engineering is a field of work and study within microbial genetics. The usage of recombinant DNA technology is a process of this work. The process involves creating recombinant DNA molecules through manipulating a DNA sequence. That DNA created is then in contact with a host organism. Cloning is also an example of genetic engineering.

Since the discovery of microorganisms by Robert Hooke and Antoni van Leeuwenhoek during the period 1665-1885 they have been used to study many processes and have had applications in various areas of study in genetics. For example: Microorganisms' rapid growth rates and short generation times are used by scientists to study evolution. Robert Hooke and Antoni van Leeuwenhoek discoveries involved depictions, observations, and descriptions of microorganisms. Mucor is the microfungus that Hooke presented and gave a depiction of. His contribution being, Mucor as the first microorganism to be illustrated. Antoni van Leeuwenhoek's contribution to the microscopic protozoa and microscopic bacteria yielded to scientific observations and descriptions. These contributions were accomplished by a simple microscope, which led to the understanding of microbes today and continues to progress scientists understanding.Microbial genetics also has applications in being able to study processes and pathways that are similar to those found in humans such as drug metabolism.

Role in Understanding Evolution

Microbial genetics can focus on Charles Darwin's work and scientists have continued to study his work and theories by the use of microbes. Specifically, Darwin's theory of natural selection is a source used. Studying evolution by using microbial genetics involves scientists looking at evolutionary balance. An example of how they may accomplish this is studying natural selection or drift of microbes. Application of this knowledge comes from looking for the presence or absence in a variety of different ways. The ways include identifying certain pathways, genes, and functions. Once the subject is observed, scientist may compare it to a sequence of a conserved gene. The process of studying microbial evolution in this way lacks the ability to give a time scale of when the evolution took place. However, by testing evolution in this way, scientist can learn the rates and outcomes of evolution. Studying the relationship between microbes and the environment is a key component to microbial genetics evolution.

Microorganisms whose Study is Encompassed by Microbial Genetics.

Bacteria

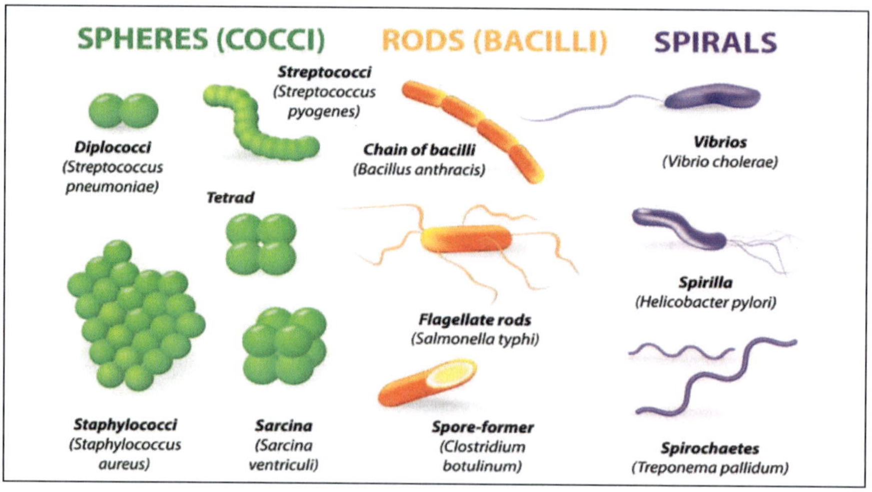

Bacteria are classified by their shape.

Bacteria have been on this planet for approximately 3.5 billion years, and are classified by their shape. Bacterial genetics studies the mechanisms of their heritable information, their chromosomes, plasmids, transposons, and phages.

Gene transfer systems that have been extensively studied in bacteria include genetic transformation, conjugation and transduction. Natural transformation is a bacterial adaptation for DNA transfer between two cells through the intervening medium. The uptake of donor DNA and its recombinational incorporation into the recipient chromosome depends on the expression of numerous bacterial genes whose products direct this process. In general, transformation is a complex, energy-requiring developmental process that appears to be an adaptation for repairing DNA damage.

Bacterial conjugation is the transfer of genetic material between bacterial cells by direct cell-to-cell contact or by a bridge-like connection between two cells. Bacterial conjugation has been extensively studied in *Escherichia coli*, but also occurs in other bacteria such as *Mycobacterium smegmatis*. Conjugation requires stable and extended contact between a donor and a recipient strain, is DNase resistant, and the transferred DNA is incorporated into the recipient chromosome by homologous recombination. *E. coli* conjugation is mediated by expression of plasmid genes, whereas mycobacterial conjugation is mediated by genes on the bacterial chromosome.

Transduction is the process by which foreign DNA is introduced into a cell by a virus or viral vector. Transduction is a common tool used by molecular biologists to stably introduce a foreign gene into a host cell's genome.

Archaea

Archaea is a domain of organisms that are prokaryotic, single-celled, and are thought to have developed 4 billion years ago. "They have no cell nucleus or any other organelles inside their cells."Archaea replicate asexually in a process known as binary fission. The cell division cycle includes when chromosomes of daughter cells replicate. Because archea have a singular structure chromosome, the two daughter cells separate and cell divides. Archaea have motility include with flagella, which is a tail like structure. Archaeal chromosomes replicate from different origins of replication, producing two haploid daughter cells. They share a common ancestor with bacteria, but are more closely related to eukaryotes in comparison to bacteria. Some Archaea are able to survive extreme environments, which leads to many applications in the field of genetics. One of such applications is the use of archaeal enzymes, which would be better able to survive harsh conditions *in vitro*.

Gene transfer and genetic exchange have been studied in the halophilic archaeon *Halobacterium volcanii* and the hyperthermophilic archaeons *Sulfolobus solfataricus* and *Sulfolobus acidocaldarius*. *H. volcani* forms cytoplasmic bridges between cells that appear to be used for transfer of DNA from one cell to another in either direction. When *S. solfataricus* and *S. acidocaldarius* are exposed to DNA damaging agents, species-specific cellular aggregation is induced. Cellular aggregation mediates chromosomal marker exchange and genetic recombination with high frequency. Cellular aggregation is thought to enhance species specific DNA transfer between *Sulfolobus* cells in order to provide increased repair of damaged DNA by means of homologous recombination. Archaea are divided into 3 subgroups which are halophiles, methanogens, and thermoacidophiles. The first group, methanogens, are archaeabacteria that live in swamps and marshes as well as in the gut of humans. They also play a major role in decay and decomposition with dead organisms. Methanogens are anaerobic organisms, which are killed when they are exposed to oxygen. The second subgroup of archaeabacteria, halophiles are organisms that are present in areas with high salt concentration like the Great Salt Lake and the Dead Sea. The third subgroup thermoacidophiles also called thermophiles, are organisms that live in acidic areas. They are present in area with low pH levels like hot springs and geyers. Most thermophiles are found in the Yellowstone National Park.

Archaeal Genetics is the study of genes that consist of single nucleus-free cells. Archaea have a single, circular chromosomes that contain multiple origins of replication for initiation of DNA synthesis. DNA replication of Archaea involves similar processes including initiation, elongation, and termination. The primase used to synthesize a RNA primer varies than in eukaryotes. The primase by archaea is highly derived version of RNA recognition motif(RRM). Archaea come from Gram positive bacteria, which both have a single lipid bilayer, which are resistant to antibiotics. Archaea are similar to mitochondria in eukaryotes in that they release energy as adenosine triphosphate (ATP) through the chemical reaction called metabolism. Some archaea known as phototrophic archaea use the sun's energy to produce ATP. ATP synthase is used as photophosphorylation to convert chemicals into ATP.

Archaea and bacteria are structurally similar even though they are not closely related in the tree of life. The shapes of both bacteria and archaea cells vary from a spherical shape known as coccus or a rod-shape known as bacillus. They are also related with no internal membrane and a cell wall that assists the cell maintaining its shape. Even though archaeal cells have cells walls, they do not contain peptidoglycan, which means archaea do not produce cellulose or chitin. Archaea are most closely related to eukaryotes due to tRNA present in archaea, but not in bacteria. Archaea have the same ribosomes as eukaryotes that synthesize into proteins. Aside from the morphology of archaea and bacteria, there are other differences between these domains. Archaea that live in extreme and harsh environments with low pH levels such as salt lakes, oceans, and in the gut of ruminants and humans are also known as extremophiles. In contrast, bacteria are found in various areas such as plants, animals, soil, and rocks.

Fungi

Fungi can be both multicellular and unicellular organisms, and are distinguished from other microbes by the way they obtain nutrients. Fungi secrete enzymes into their surroundings, to break down organic matter. Fungal genetics uses yeast, and filamentous fungi as model organisms for eukaryotic genetic research, including cell cycle regulation, chromatin structure and gene regulation.

Studies of the fungus *Neurospora crassa* have contributed substantially to understanding how genes work. *N. crassa* is a type of red bread mold of the phylum *Ascomycota*. It is used as a model organism because it is easy to grow and has a haploid life cycle that makes genetic analysis simple since recessive traits will show up in the offspring. Analysis of genetic recombination is facilitated by the ordered arrangement of the products of meiosis in ascospores. In its natural environment, *N. crassa* lives mainly in tropical and sub-tropical regions. It often can be found growing on dead plant matter after fires.

Neurospora was used by Edward Tatum and George Beadle in their experiments for which they won the Nobel Prize in Physiology or Medicine in 1958. The results of these experiments led directly to the one gene-one enzyme hypothesis that specific genes code for specific proteins. This concept proved to be the opening gun in what became molecular genetics and all the developments that have followed from that.

Saccharomyces cerevisiae is a yeast of the phylum *Ascomycota*. During vegetative growth that ordinarily occurs when nutrients are abundant, *S. cerevisiae* reproduces by mitosis as diploid cells. However, when starved, these cells undergo meiosis to form haploid spores. Mating occurs when haploid cells of opposite mating types MATa and MATα come into contact. Ruderfer et al. pointed out that, in nature, such contacts are frequent between closely related yeast cells for two reasons. The first is that cells of opposite mating type are present together in the same acus, the sac that contains the cells directly produced by a single meiosis, and these cells can mate with each other. The second reason is that haploid cells of one mating type, upon cell division, often produce cells of the opposite mating type. An analysis of the ancestry of natural *S. cerevisiae* strains concluded that outcrossing occurs very infrequently (only about once every 50,000 cell divisions). The relative rarity in nature of meiotic events that result from outcrossing suggests that the possible long-term benefits of outcrossing (e.g. generation of diversity) are unlikely to be sufficient for generally maintaining sex from one generation to the next. Rather, a short term benefit, such as meiotic recombinational repair of DNA damages caused by stressful conditions (such as starvation) may be the key to the maintenance of sex in *S. cerevisiae*.

Candida albicans is a diploid fungus that grows both as a yeast and as a filament. C. albicans is the most common fungal pathogen in humans. It causes both debilitating mucosal infections and potentially life-threatening systemic infections. *C. albicans* has maintained an elaborate, but largely hidden, mating apparatus. Johnson suggested that mating strategies may allow *C. albicans* to survive in the hostile environment of a mammalian host.

Among the 250 known species of aspergilli, about 33% have an identified sexual state. Among those *Aspergillus* species that exhibit a sexual cycle the overwhelming majority in nature are homothallic (self-fertilizing). Selfing in the homothallic fungus *Aspergillus nidulans* involves activation of the same mating pathways characteristic of sex in outcrossing species, i.e. self-fertilization does not bypass required pathways for outcrossing sex but instead requires activation of these pathways within a single individual. Fusion of haploid nuclei occurs within reproductive structures termed *cleistothecia*, in which the diploid zygote undergoes meiotic divisions to yield haploid ascospores.

Protozoa

Protozoa are unicellular organisms, which have nuclei, and ultramicroscopic cellular bodies within their cytoplasm. One particular aspect of protozoa that are of interest to human geneticists are their flagella, which are very similar to human sperm flagella.

Studies of *Paramecium* have contributed to our understanding of the function of meiosis. Like all ciliates, *Paramecium* has a polyploid macronucleus, and one or more diploid micronuclei. The macronucleus controls non-reproductive cell functions, expressing the genes needed for daily functioning. The micronucleus is the generative, or germline nucleus, containing the genetic material that is passed along from one generation to the next.

In the asexual fission phase of growth, during which cell divisions occur by mitosis rather than meiosis, clonal aging occurs leading to a gradual loss of vitality. In some species, such as the well studied *Paramecium tetraurelia*, the asexual line of clonally aging paramecia loses vitality and expires after about 200 fissions if the cells fail to undergo meiosis followed by either autogamy (self-fertilizaion) or conjugation (outcrossing). DNA damage increases dramatically during successive clonal cell divisions and is a likely cause of clonal aging in *P. tetraurelia*.

When clonally aged *P. tetraurelia* are stimulated to undergo meiosis in association with either autogamy or conjugation, the progeny are rejuvenated, and are able to have many more mitotic binary fission divisions. During either of these processes the micronuclei of the cell(s) undergo meiosis, the old macronucleus disintegrates and a new macronucleus is formed by replication of the micronuclear DNA that had recently undergone meiosis. There is apparently little, if any, DNA damage in the new macronucleus, suggesting that rejuvenation is associated with the repair of these damages in the micronucleus during meiosis.

Viruses

Viruses are capsid-encoding organisms composed of proteins and nucleic acids that can self-assemble after replication in a host cell using the host's replication machinery. There is a disagreement in science about whether viruses are living due to their lack of ribosomes. Comprehending the viral genome is important not only for studies in genetics but also for understanding their pathogenic properties.

Many types of virus are capable of genetic recombination. When two or more individual viruses of the same type infect a cell, their genomes may recombine with each other to produce recombinant virus progeny. Both DNA and RNA viruses can undergo recombination. When two or more viruses, each containing lethal genomic damage infect the same host cell, the virus genomes often can pair with each other and undergo homologous recombinational repair to produce viable progeny. This process is known as multiplicity reactivation. Enzymes employed in multiplicity reactivation are functionally homologous to enzymes employed in bacterial and eukaryotic recombinational repair. Multiplicity reactivation has been found to occur with pathogenic viruses including influenza virus, HIV-1, adenovirus simian virus 40, vaccinia virus, reovirus, poliovirus and herpes simplex virus as well as numerous bacteriophages.

Any living organism can contract a virus by giving parasites the opportunity to grow. Parasites feed on the nutrients of another organism which allows the virus to thrive. Once the human body detects a virus, it then creates fighter cells that attack the parasite/virus; literally, causing a war within the body. A virus can affect any part of the body causing a wide range of illnesses such as the flu, the common cold, and sexually transmitted diseases. The flu is an airborne virus that travels through tiny droplets and is formally known as Influenza. Parasites travel through the air and attack the human respiratory system. People that are initially infected with this virus pass infection on by normal day to day activity such as talking and sneezing. When a person comes in contact with the virus, unlike the common cold, the flu virus affects people almost immediately. Symptoms of this virus are very similar to the common cold but much worse. Body aches, sore throat, headache, cold sweats, muscle aches and fatigue are among the many symptoms accompanied by the virus. A viral infection in the upper respiratory tract results in the common cold. With symptoms like sore throat, sneezing, small fever, and a cough, the common cold is usually harmless and tends to clear up within a week or so. The common cold is also a virus that is spread through the air but can also be passed through direct contact. This infection takes a few days to develop symptoms; it is a gradual process unlike the flu.

Applications of Microbial Genetics

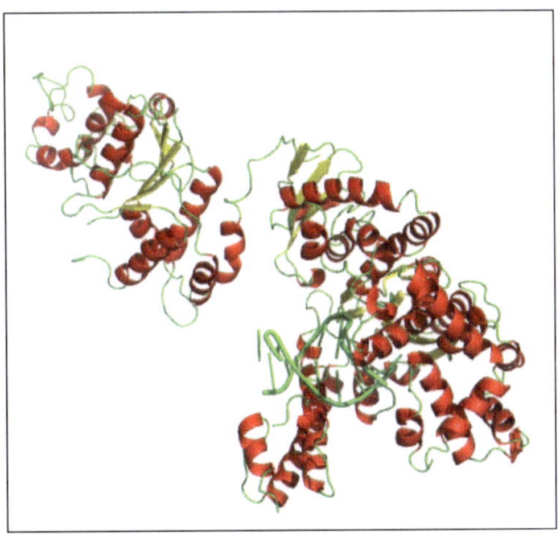

Taq polymerase which is used in Polymerase Chain Reaction (PCR).

Microbes are ideally suited for biochemical and genetics studies and have made huge contributions to these fields of science such as the demonstration that DNA is the genetic material, that the gene has a simple linear structure, that the genetic code is a triplet code, and that gene expression is regulated by specific genetic processes. Jacques Monod and François Jacob used *Escherichia coli*, a type of bacteria, in order to develop the operon model of gene expression, which lay down the basis of gene expression and regulation. Furthermore, the hereditary processes of single-celled eukaryotic microorganisms are similar to those in multi-cellular organisms allowing researchers to gather information on this process as well. Another bacterium which has greatly contributed to the field of genetics is *Thermus aquaticus*, which is a bacterium that tolerates high temperatures. From this microbe scientists isolated the enzyme Taq polymerase, which is now used in the powerful experimental technique, Polymerase chain reaction(PCR). Additionally the development of recombinant DNA technology through the use of bacteria has led to the birth of modern genetic engineering and biotechnology.

Using microbes, protocols were developed to insert genes into bacterial plasmids, taking advantage of their fast reproduction, to make biofactories for the gene of interest. Such genetically engineered bacteria can produce pharmaceuticals such as insulin, human growth hormone, interferons and blood clotting factors. These biofactories are typically much cheaper to operate and maintain than the alternative procedures of producing pharmaceuticals. They're like millions of tiny pharmaceutical machines that only require basic raw materials and the right environment to produce a large amount of product. The utilization of incorporating the human insulin gene alone has had profound impacts on the medical industry. It's thought that biofactories might be the ultimate key in reducing the price of expensive life saving pharmaceutical compounds.

Microbes synthesize a variety of enzymes for industrial applications, such as fermented foods, laboratory test reagents, dairy products (such as renin), and even in clothing (such as *Trichoderma* fungus whose enzyme is used to give jeans a stone washed appearance).

There is currently potential for microbes to be used as an alternative for petroleum based surfactants. Microbial surfactants would still have the same kind of hydrophillic and hydrophobic functional groups as their petroleum based counterparts, but they have numerous advantages over their competition. In comparison, microbial amphiphillic compounds have robust a tendency to stay functional in extreme environments such as areas with high heat or extreme ph. all while being biodegradable and less toxic to the environment. This efficient and cheap method of production could be the solution to the ever increasing global consumption of surfactants. Ironically, the application for bio-based surfactants with the most demand is the oil industry which uses surfactants in general production as well as development of specific oil compositions.

Microbes are an abundant source of lipases which have a wide variety of industrial and consumer applications. Enzymes perform a wide variety of functions inside the cells of living things, so it only makes sense that we can use them for similar purposes on a larger scale. Microbial enzymes are typically preferred for mass production due to the wide variety of functions available and their ability to be mass produced. Plant and animal enzymes are typically too expensive to be mass produced, however this is not always the case. Especially in plants. Industrial applications of lipases generally include the enzyme as a more efficient and cost effective catalyst in the production of commercially valuable chemicals from fats and oils, because they are able to retain their specific properties in mild easy to maintain conditions and work at an increased rate. Other already successful applications of lipolytic enzymes include the production of biofuels, polymers, non-stereoisomeric pharmaceuticals, agricultural compounds, and flavor enhancing compounds.

In regards to industrial optimization, the benefit of the biofactory method of production is the ability to direct optimization by means of directed evolution. The efficiency and specificity of production will increase over time by imposing artificial selection. This method of improving efficiency is nothing new in the agriculture, but it's a relatively new concept in industrial production. It is though that this method will be far superior to conventional industrial methods because you have optimization on multiple fronts. The first front being that the microorganisms that make up biofactories can be evolved to our needs. The second front being the conventional method of optimization brought about by the integration of advancing technologies. This combination of conventional and biological advancement is just now becoming utilized and provides a virtually limitless number of applications.

Developmental Genetics

Developmental genetics is the study of how genes control the growth and development of an organism throughout its life-cycle.

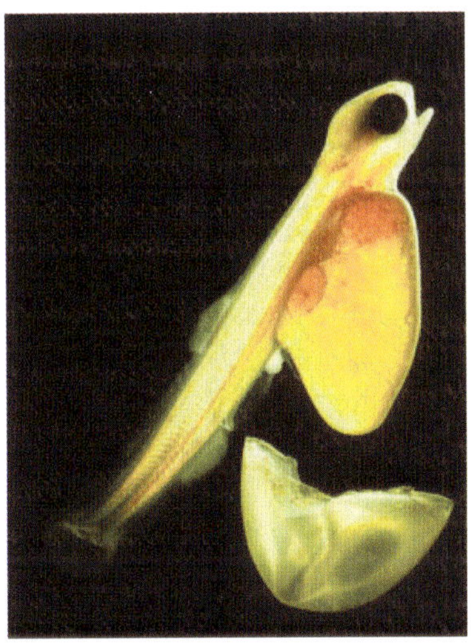

Genes code for proteins, and proteins build bodies: a salmon fry hatching from an egg.

The function of genes is to pass on the information necessary to build proteins - and bodies - from one generation to the next. A newly fertilised egg cell has a collection of genes that contains all information needed to transform it from a single cell into an embryo and then an adult. The process that changes a single cell into a new person (or a new frog, or a new oak tree) is called development.

During the course of development, complex structures develop from simple ones. A single cell transforms itself into an adult organism. How does something complicated come from something simple? And how do genes control this process?

Creating an organism from a single cell involves three important processes:

- Cell division: cells divide to produce more cells.
- Cell differentiation: cells change into different types of cell to do specific jobs in the body, from nerve cells to muscle cells.
- Morphogenesis: groups of cells move and change their shape to produce the structure of the organism.

Genes play a vital role in controlling all of these processes.

Genes contain the information a cell needs to make proteins - a bit like a recipe for a living thing. Different genes contain the information needed to make different proteins, and different proteins do different jobs in the cell. The proteins a cell makes decide what kind of cell it becomes, and there are some 350 different types of cell in an adult human being.

Cells change into different types of cell because of changes in the way their genes work. Some genes are activated (switched on), and some genes are inactivated (switched off). As a result, the cell produces a specific set of proteins. So, for example, a nerve cell produces only the proteins needed to make a nerve cell, and a muscle cell produces only the proteins needed to make a muscle cell.

But how do cells switch their genes on and off? And, more importantly, how do they 'know' which genes to switch on and which genes to switch off? The answer lies in special control genes that produce proteins that control the activity of other genes. So, for example, homeotic or homeobox genes control whole sets of other genes to set out the basic body plan of the embryo, separating the front from the back, and producing the right body structure in the right place.

One way in which genes can influence the activity of other genes is through the production of proteins called transcription factors, which stick to special control sites in the DNA at the start of a gene to switch them on and off.

Cell Division

In unicellular organisms, cell division is the means of reproduction; in multicellular organisms, it is the means of tissue growth and maintenance. Survival of the eukaryotes depends upon interactions between many cell types, and it is essential that a balanced distribution of types be maintained. This is achieved by the highly regulated process of cell proliferation. The growth and division of different cell populations are regulated in different ways, but the basic mechanisms are similar throughout multicellular organisms.

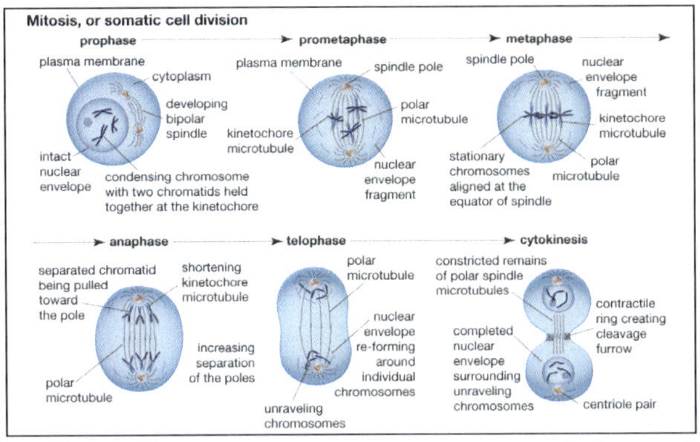

Mitosis: One cell gives rise to two genetically identical daughter cells during the process of mitosis.

Most tissues of the body grow by increasing their cell number, but this growth is highly regulated to maintain a balance between different tissues. In adults most cell division is involved in tissue renewal rather than growth, many types of cells undergoing continuous replacement. Skin cells, for example, are constantly being sloughed off and replaced; in this case, the mature differentiated cells do not divide, but their population is renewed by division of immature stem cells. In certain other cells, such as those of the liver, mature cells remain capable of division to allow growth or regeneration after injury.

In contrast to these patterns, other types of cells either cannot divide or are prevented from dividing by certain molecules produced by nearby cells. As a result, in the adult organism, some tissues have a greatly reduced capacity to renew damaged or diseased cells. Examples of such tissues include heart muscle, nerve cells of the central nervous system, and lens cells in mammals. Maintenance and repair of these cells is limited to replacing intracellular components rather than replacing entire cells.

Duplication of the Genetic Material

Before a cell can divide, it must accurately and completely duplicate the genetic information encoded in its DNA in order for its progeny cells to function and survive. This is a complex problem because of the great length of DNA molecules. Each human chromosome consists of a long double spiral, or helix, each strand of which consists of more than 100 million nucleotides.

The duplication of DNA is called DNA replication, and it is initiated by complex enzymes called DNA polymerases. These progress along the molecule, reading the sequences of nucleotides that are linked together to make DNA chains. Each strand of the DNA double helix, therefore, acts as a template specifying the nucleotide structure of a new growing chain. After replication, each of the two daughter DNA double helices consists of one parental DNA strand wound around one newly synthesized DNA strand.

In order for DNA to replicate, the two strands must be unwound from each other. Enzymes called helicases unwind the two DNA strands, and additional proteins bind to the separated strands to stabilize them and prevent them from pairing again. In addition, a remarkable class of enzyme called DNA topoisomerase removes the helical twists by cutting either one or both strands and then resealing the cut. These enzymes can also untangle and unknot DNA when it is tightly coiled into a chromatin fibre.

In the circular DNA of prokaryotes, replication starts at a unique site called the origin of replication and then proceeds in both directions around the molecule until the two processes meet, producing two daughter molecules. In rapidly growing prokaryotes, a second round of replication can start before the first has finished. The situation in eukaryotes is more complicated, as replication moves more slowly than in prokaryotes. At 500 to 5,000 nucleotides per minute (versus 100,000 nucleotides per minute in prokaryotes), it would take a human chromosome about a month to replicate if started at a single site. Actually, replication begins at many sites on the long chromosomes of animals, plants, and fungi. Distances between adjacent initiation sites are not always the same; for example, they are closer in the rapidly dividing embryonic cells of frogs or flies than in adult cells of the same species.

Accurate DNA replication is crucial to ensure that daughter cells have exact copies of the genetic information for synthesizing proteins. Accuracy is achieved by a "proofreading" ability of the DNA polymerase itself. It can erase its own errors and then synthesize anew. There are also repair systems that correct genetic damage to DNA. For example, the incorporation of an incorrect nucleotide, or damage caused by mutagenic agents, can be corrected by cutting out a section of the daughter strand and recopying the parental strand.

Cell Differentiation

Adult organisms are composed of a number of distinct cell types. Cells are organized into tissues, each of which typically contains a small number of cell types and is devoted to a specific physiological function. For example, the epithelial tissue lining the small intestine contains columnar absorptive cells, mucus-secreting goblet cells, hormone-secreting endocrine cells, and enzyme-secreting Paneth cells. In addition, there exist undifferentiated dividing cells that lie in the crypts between the intestinal villi and serve to replace the other cell types when they become damaged or worn out. Another example of a differentiated tissue is the skeletal tissue of a long bone, which contains osteoblasts (large cells that synthesize bone) in the outer sheath and osteocytes (mature bone cells) and osteoclasts (multinucleate cells involved in bone remodeling) within the matrix.

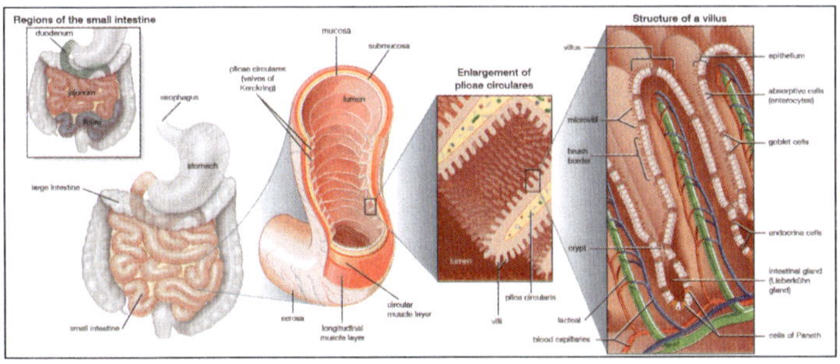

The small intestine contains many distinct types of cells, each of which serves a specific function.

In general, the simpler the overall organization of the animal, the fewer the number of distinct cell types that they possess. Mammals contain more than 200 different cell types, whereas simple invertebrate animals may have only a few different types. Plants are also made up of differentiated cells, but they are quite different from the cells of animals. For example, a leaf in a higher plant is covered with a cuticle layer of epidermal cells. Among these are pores composed of two specialized cells, which regulate gaseous exchange across the epidermis. Within the leaf is the mesophyll, a spongy tissue responsible for photosynthetic activity. There are also veins composed of xylem elements, which transport water up from the soil, and phloem elements, which transport products of photosynthesis to the storage organs.

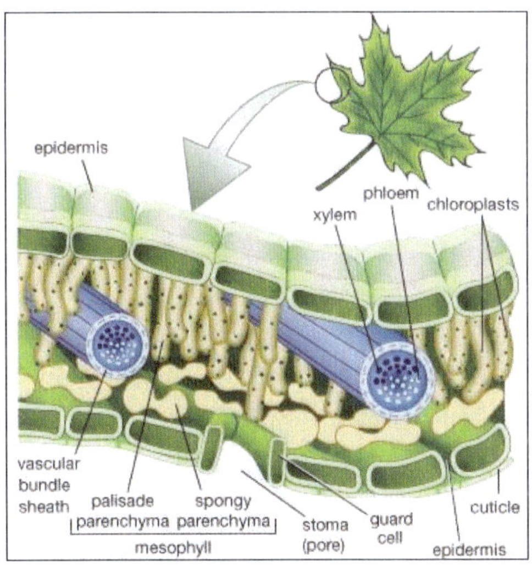

Structures of a leaf

The epidermis is often covered with a waxy protective cuticle that helps prevent water loss from inside the leaf. Oxygen, carbon dioxide, and water enter and exit the leaf through pores (stomata) scattered mostly along the lower epidermis. The stomata are opened and closed by the contraction and expansion of surrounding guard cells. The vascular, or conducting, tissues are known as xylem and phloem; water and minerals travel up to the leaves from the roots through the xylem, and sugars made by photosynthesis are transported to other parts of the plant through the phloem. Photosynthesis occurs within the chloroplast-containing mesophyll layer.

The various cell types have traditionally been recognized and classified according to their appearance in the light microscope following the process of fixing, processing, sectioning, and staining tissues that is known as histology. Classical histology has been augmented by a variety of more discriminating techniques. Electron microscopy allows for higher magnifications. Histochemistry involves the use of coloured precipitating substrates to stain particular enzymes in situ. Immunohistochemistry uses specific antibodies to identify particular substances, usually proteins or carbohydrates, within cells. In situ hybridization involves the use of nucleic acid probes to visualize the location of specific messenger RNAs (mRNA). These modern methods have allowed the identification of more cell types than could be visualized by classical histology, particularly in the brain, the immune system, and among the hormone-secreting cells of the endocrine system.

Differentiated State

The biochemical basis of cell differentiation is the synthesis by the cell of a particular set of proteins, carbohydrates, and lipids. This synthesis is catalyzed by proteins called enzymes. Each enzyme in turn is synthesized in accordance with a particular gene, or sequence of nucleotides in the DNA of the cell nucleus. A particular state of differentiation, then, corresponds to the set of genes that is expressed and the level to which it is expressed.

It is believed that all of an organism's genes are present in each cell nucleus, no matter what the cell type, and that differences between tissues are not due to the presence or absence of certain genes but are due to the expression of some and the repression of others. In animals the best evidence for retention of the entire set of genes comes from whole animal cloning experiments in which the nucleus of a differentiated cell is substituted for the nucleus of a fertilized egg. In many species this can result in the development of a normal embryo that contains the full range of body parts and cell types. Likewise, in plants it is often possible to grow complete embryos from individual cells in tissue culture. Such experiments show that any nucleus has the genetic information required for the growth of a developing organism, and they strongly suggest that, for most tissues, cell differentiation arises from the regulation of genetic activity rather than the removal or destruction of unwanted genes. The only known exception to this rule comes from the immune system, where segments of DNA in developing white blood cells are slightly rearranged, producing a wide variety of antibody and receptor molecules.

At the molecular level there are many ways in which the expression of a gene can be differentially regulated in different cell types. There may be differences in the copying, or transcription, of the gene into RNA; in the processing of the initial RNA transcript into mRNA; in the control of mRNA movement to the cytoplasm; in the translation of mRNA to protein; or in the stability of mRNA. However, the control of transcription has the most influence over gene expression and has received the most detailed analysis.

The DNA in the cell nucleus exists in the form of chromatin, which is made up of DNA bound to histones (simple alkaline proteins) and other nonhistone proteins. Most of the DNA is complexed into repeating structures called nucleosomes, each of which contains eight molecules of histone. Active genes are found in parts of the DNA where the chromatin has an "open" configuration, in which regulatory proteins are able to gain access to the DNA. The degree to which the chromatin opens depends on chemical modifications of the outer parts of the histone molecules and on the presence or absence of particular nonhistone proteins. Transcriptional control is exerted with the help of regulatory sequences that are found associated with a gene, such as the promoter sequence, a region near the start of the gene, and enhancer sequences, regions that lie elsewhere within the DNA that augment the activity of enzymes involved in the process of transcription. Whether or not transcription occurs depends on the binding of transcription factors to these regulatory sequences. Transcription factors are proteins that usually possess a DNA-binding region, which recognizes the specific regulatory sequence in the DNA, and an effector region, which activates or inhibits transcription. Transcription factors often work by recruiting enzymes that add modifications (e.g., acetyl groups or methyl groups) to or remove modifications from the outer parts of the histone molecules. This controls the folding of the chromatin and the accessibility of the DNA to RNA polymerase and other transcription factors.

In general, it requires several transcription factors working in combination to activate a gene. For example, the chicken delta 1 crystallin gene, normally expressed only in the lens of the eye, has a promoter that contains binding sites for two activating transcription factors and an enhancer that contains binding sites for two other activating transcription factors. There is also an additional enhancer site that can bind either an activator (deltaEF3) or a repressor (deltaEF1). Successful transcription requires that all these sites are occupied by the correct transcription factors.

Fully differentiated cells are qualitatively different from one another. States of terminal differentiation are stable and persistent, both in the lifetime of the cell and in successive cell generations (in the case of differentiated types that are capable of continued cell division). The inherent stability of the differentiated state is maintained by various processes, including feedback activation of genes by their own products and repression of inactive genes. Chromatin structure may be important in maintaining states of differentiation, although it is still unclear whether this can be maintained during DNA replication, which involves temporary removal of chromosomal proteins and unwinding of the DNA double helix.

A type of differentiation control that is maintained during DNA replication is the methylation of DNA, which tends to recruit histone deacetylases and hence close up the structure of the chromatin. DNA methylation occurs when a methyl group is attached to the exterior, or sugar-phosphate side, of a cytosine (C) residue. Cytosine methylation occurs only on a C nucleotide when it is connected to a G (guanine) nucleotide on the same strand of DNA. These nucleotide pairings are called CG dinucleotides. One class of DNA methylase enzyme can introduce new methylations when required, whereas another class, called maintenance methylases, methylates CG dinucleotides in the DNA double helix only when the CG of the complementary strand is already methylated. Each time the methylated DNA is replicated, the old strand has the methyl groups and the new strand does not. The maintenance methylase will then add methyl groups to all the CGs opposite the existing methyl groups to restore a fully methylated double helix. This mechanism guarantees stability of the DNA methylation pattern, and hence the differentiated state, during the processes of DNA replication and cell division.

The Process of Differentiation

Differentiation from visibly undifferentiated precursor cells occurs during embryonic development, during metamorphosis of larval forms, and following the separation of parts in asexual reproduction. It also takes place in adult organisms during the renewal of tissues and the regeneration of missing parts. Thus, cell differentiation is an essential and ongoing process at all stages of life.

The visible differentiation of cells is only the last of a progressive sequence of states. In each state, the cell becomes increasingly committed toward one type of cell into which it can develop. States of commitment are sometimes described as "specification" to represent a reversible type of commitment and as "determination" to represent an irreversible commitment. Although states of specification and determination both represent differential gene activity, the properties of embryonic cells are not necessarily the same as those of fully differentiated cells. In particular, cells in specification states are usually not stable over prolonged periods of time.

Two mechanisms bring about altered commitments in the different regions of the early embryo: cytoplasmic localization and induction. Cytoplasmic localization is evident in the earliest stages of development of the embryo. During this time, the embryo divides without growth, undergoing cleavage divisions that produce separate cells called blastomeres. Each blastomere inherits a certain region of the original egg cytoplasm, which may contain one or more regulatory substances called cytoplasmic determinants. When the embryo has become a solid mass of blastomeres (called a morula), it generally consists of two or more differently committed cell populations—a result of the blastomeres having incorporated different cytoplasmic determinants. Cytoplasmic determinants may consist of mRNA or protein in a particular state of activation. An example of the influence of a cytoplasmic determinant is a receptor called Toll, located in the membranes of Drosophila (fruit fly) eggs. Activation of Toll ensures that the blastomeres will develop into ventral (underside) structures, while blastomeres containing inactive Toll will produce cells that will develop into dorsal (back) structures.

In induction, the second mechanism of commitment, a substance secreted by one group of cells alters the development of another group. In early development, induction is usually instructive; that is, the tissue assumes a different state of commitment in the presence of the signal than it would in the absence of the signal. Inductive signals often take the form of concentration gradients of substances that evoke a number of different responses at different concentrations. This leads to the formation of a sequence of groups of cells, each in a different state of specification. For example, in Xenopus (clawed frog) the early embryo contains a signaling centre called the organizer that secretes inhibitors of bone morphogenetic proteins (BMPs), leading to a ventral-to-dorsal (belly-to-back) gradient of BMP activity. The activity of BMP in the ventral region of the embryo suppresses the expression of transcription factors involved in the formation of the central nervous system and segmented muscles. Suppression ensures that these structures are formed only on the dorsal side, where there is decreased activity of BMP.

The final stage of differentiation often involves the formation of several types of differentiated cells from one precursor or stem cell population. Terminal differentiation occurs not only in embryonic development but also in many tissues in postnatal life. Control of this process depends on a system of lateral inhibition in which cells that are differentiating along a particular pathway send out signals that repress similar differentiation by their neighbours. For example, in the developing central nervous system of vertebrates, neurons arise from a simple tube of neuroepithelium, the cells of which possess a surface receptor called Notch. These cells also possess another cell surface molecule called Delta that can bind to and activate Notch on adjacent cells. Activation of Notch initiates a cascade of intracellular events that results in suppression of Delta production and suppression of neuronal differentiation. This means that the neuroepithelium generates only a few cells with high expression of Delta surrounded by a larger number of cells with low expression of Delta. High Delta production and low Notch activation makes the cells develop into neurons. Low Delta production and high Notch activation makes the cells remain as precursor cells or become glial (supporting) cells. A similar mechanism is known to produce the endocrine cells of the pancreas and the goblet cells of the intestinal epithelium. Such lateral inhibition systems work because cells in a population are never quite identical to begin with. There are always small differences, such as in the number of Delta molecules displayed on the cell surface. The mechanism of lateral inhibition amplifies these small differences, using them to bring about differential gene expression that leads to stable and persistent states of cell differentiation.

Errors in Differentiation

Three classes of abnormal cell differentiation are dysplasia, metaplasia, and anaplasia. Dysplasia indicates an abnormal arrangement of cells, usually arising from a disturbance in their normal growth behaviour. Some dysplasias are precursor lesions to cancer, whereas others are harmless and regress spontaneously. For example, dysplasia of the uterine cervix, called cervical intraepithelial neoplasia (CIN), may progress to cervical cancer. It can be detected by cervical smear cytology tests (Pap smears).

Metaplasia is the conversion of one cell type into another. In fact, it is not usually the differentiated cells themselves that change but rather the stem cell population from which they are derived. Metaplasia commonly occurs where chronic tissue damage is followed by extensive regeneration. For example, squamous metaplasia of the bronchi occurs when the ciliated respiratory epithelial cells of people who smoke develop into squamous, or flattened, cells. In intestinal metaplasia of the stomach, patches resembling intestinal tissue arise in the gastric mucosa, often in association with gastric ulcers. Both of these types of metaplasia may progress to cancer.

Anaplasia is a loss of visible differentiation that can occur in advanced cancer. In general, early cancers resemble their tissue of origin and are described and classified by their pattern of differentiation. However, as they develop, they produce variants of more abnormal appearance and increased malignancy. Finally, a highly anaplastic growth can occur, in which the cancerous cells bear no visible relation to the parent tissue.

Evolution of Cells

Development of Genetic Information

Life on Earth could not exist until a collection of catalysts appeared that could promote the synthesis of more catalysts of the same kind. Early stages in the evolutionary pathway of cells presumably centred on RNA molecules, which not only present specific catalytic surfaces but also contain the potential for their own duplication through the formation of a complementary RNA molecule. It is assumed that a small RNA molecule eventually appeared that was able to catalyze its own duplication.

Imperfections in primitive RNA replication likely gave rise to many variant autocatalytic RNA molecules. Molecules of RNA that acquired variations that increased the speed or the fidelity of self-replication would have outmultiplied other, less-competent RNA molecules. In addition, other small RNA molecules that existed in symbiosis with autocatalytic RNA molecules underwent natural selection for their ability to catalyze useful secondary reactions such as the production of better precursor molecules. In this way, sophisticated families of RNA catalysts could have evolved together, since cooperation between different molecules produced a system that was much more effective at self-replication than a collection of individual RNA catalysts.

Another major step in the evolution of the cell would have been the development, in one family of self-replicating RNA, of a primitive mechanism of protein synthesis. Protein molecules cannot provide the information for the synthesis of other protein molecules like themselves. This information must ultimately be derived from a nucleic acid sequence. Protein synthesis is much more complex than RNA synthesis, and it could not have arisen before a group of powerful RNA catalysts evolved. Each of these catalysts presumably has its counterpart among the RNA molecules that function in the current cell:

1. There was an information RNA molecule, much like messenger RNA (mRNA), whose nucleotide sequence was read to create an amino acid sequence;

2. There was a group of adaptor RNA molecules, much like transfer RNA (tRNA), that could bind to both mRNA and a specific activated amino acid; and

3. Finally, there was an RNA catalyst, much like ribosomal RNA (rRNA), that facilitated the joining together of the amino acids aligned on the mRNA by the adaptor RNA.

At some point in the evolution of biological catalysts, the first cell was formed. This would have required the partitioning of the primitive biological catalysts into individual units, each surrounded by a membrane. Membrane formation might have occurred quite simply, since many amphiphilic molecules—half hydrophobic (water-repelling) and half hydrophilic (water-loving)—aggregate to form bilayer sheets in which the hydrophobic portions of the molecules line up in rows to form the interior of the sheet and leave the hydrophilic portions to face the water. Such bilayer sheets can spontaneously close up to form the walls of small, spherical vesicles, as can the phospholipid bilayer membranes of present-day cells.

Both stearic acid (a fatty acid) and phosphatidylcholine (a phospholipid) are composed of chemical groups that form polar "heads" and nonpolar "tails." The polar heads are hydrophilic, or soluble in water, whereas the nonpolar tails are hydrophobic, or insoluble in water. Lipid molecules of this composition spontaneously form aggregate structures such as micelles and lipid bilayers, with their hydrophilic ends oriented toward the watery medium and their hydrophobic ends shielded from the water.

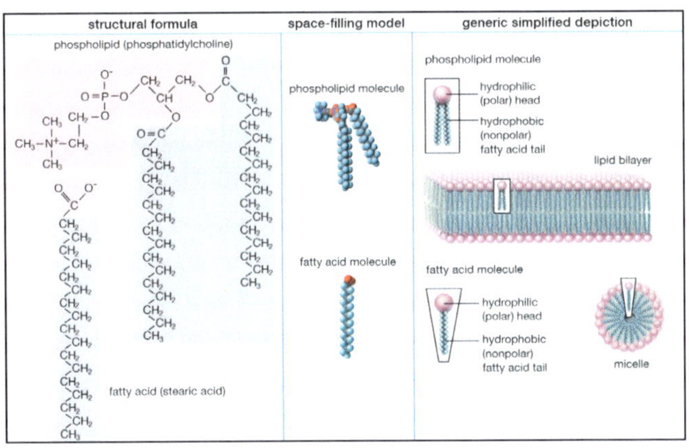

Structure and properties of two representative lipids.

As soon as the biological catalysts became compartmentalized into small individual units, or cells, the units would have begun to compete with one another for the same resources. The active competition that ensued must have greatly accelerated evolutionary change, serving as a powerful force for the development of more efficient cells. In this way, cells eventually arose that contained new catalysts, enabling them to use simpler, more abundant precursor molecules for their growth. Because these cells were no longer dependent on preformed ingredients for their survival, they were able to spread far beyond the limited environments where the first primitive cells arose.

It is often assumed that the first cells appeared only after the development of a primitive form of protein synthesis. However, it is by no means certain that cells cannot exist without proteins, and it has been suggested that the first cells contained only RNA catalysts. In either case, protein molecules, with their chemically varied side chains, are more powerful catalysts than RNA molecules; therefore, as time passed, cells arose in which RNA served primarily as genetic material, being directly replicated in each generation and inherited by all progeny cells in order to specify proteins.

As cells became more complex, a need would have arisen for a stabler form of genetic information storage than that provided by RNA. DNA, related to RNA yet chemically stabler, probably appeared rather late in the evolutionary history of cells. Over a period of time, the genetic information in RNA sequences was transferred to DNA sequences, and the ability of RNA molecules to replicate directly was lost. It was only at this point that the central process of biology—the synthesis, one after the other, of DNA, RNA, and protein—appeared.

Morphogenesis

Morphogenesis is the shaping of an organism by embryological processes of differentiation of cells, tissues, and organs and the development of organ systems according to the genetic "blueprint" of the potential organism and environmental conditions.

Plant morphogenesis is brought about chiefly through differential growth. Permanent embryonic tissue results in a morphogenetic potential that varies greatly with the environment and continues to produce new organs throughout the life of the plant. Animal morphogenesis is accomplished by growth and by cell movement. A fixed pattern is established early; the organism is determined as to shape, size, and organ complement. Once organs are formed, no new ones (with few exceptions) are produced.

Genetic and Molecular Basis

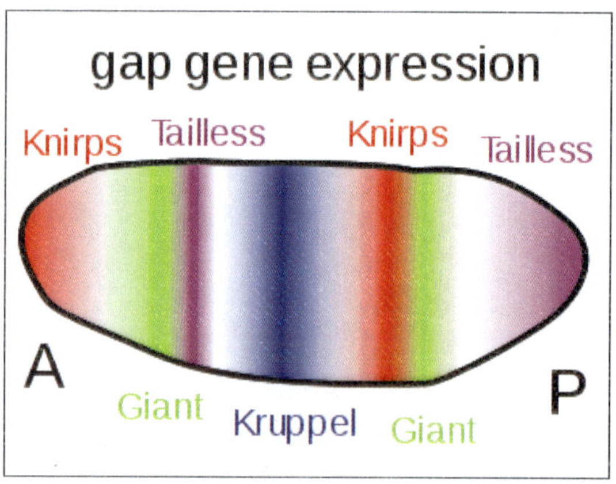

Morphogenesis is controlled by a "toolkit" of genes which switch development on and off at precise times and places. Here, gap genes in the fruit fly are switched on by genes such as bicoid, setting up stripes which create the body's segmental form.

Several types of molecules are important in morphogenesis. Morphogens are soluble molecules that can diffuse and carry signals that control cell differentiation via concentration gradients. Morphogens typically act through binding to specific protein receptors. An important class of molecules involved in morphogenesis are transcription factor proteins that determine the fate of cells by interacting with DNA. These can be coded for by master regulatory genes, and either activate or deactivate the transcription of other genes; in turn, these secondary gene products can regulate the expression of still other genes in a regulatory cascade of gene regulatory networks. At the end of this cascade are classes of molecules that control cellular behaviors such as cell migration, or, more generally, their properties, such as cell adhesion or cell contractility. For example, during gastrulation, clumps of stem cells switch off their cell-to-cell adhesion, become migratory, and take up new positions within an embryo where they again activate specific cell adhesion proteins and form new tissues and organs. Developmental signaling pathways implicated in morphogenesis include Wnt, Hedgehog, and ephrins.

Cellular Basis

At a tissue level, ignoring the means of control, morphogenesis arises because of cellular proliferation and motility. Morphogenesis also involves changes in the cellular structure or how cells interact in tissues. These changes can result in tissue elongation, thinning, folding, invasion or separation of one tissue into distinct layers. The latter case is often referred as cell sorting. Cell "sorting out" consists of cells moving so as to sort into clusters that maximize contact between cells of the same type. The ability of cells to do this has been proposed to arise from differential cell adhesion by Malcolm Steinberg through his Differential Adhesion Hypothesis. Tissue separation can also occur via more dramatic cellular differentiation events during which epithelial cells become mesenchymal. Mesenchymal cells typically leave the epithelial tissue as a consequence of changes in cell adhesive and contractile properties. Following epithelial-mesenchymal transition, cells can migrate away from an epithelium and then associate with other similar cells in a new location.

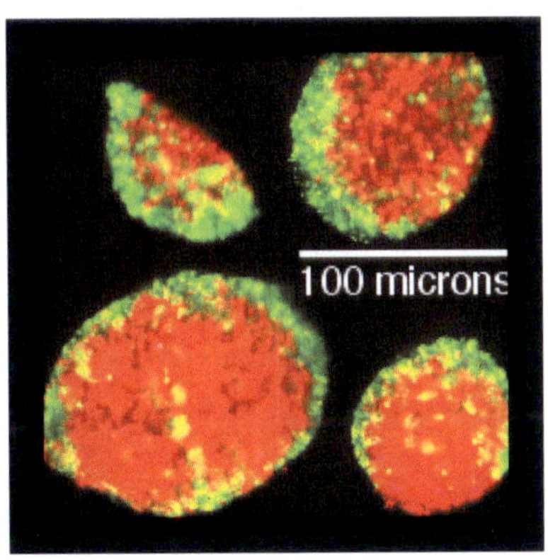

Cell sorting out with cultured P19 embryonal carcinoma cells. Live cells were stained with DiI (red) or DiO (green). The red cells were genetically altered and express higher levels of E-cadherin than the green cells. The mixed culture forms large multi-cellular aggregates.

Cell-cell Adhesion

During embryonic development, cells are restricted to different layers due to differential affinities. One of the ways this can occur is when cells share the same cell-to-cell adhesion molecules. For instance, homotypic cell adhesion can maintain boundaries between groups of cells that have different adhesion molecules. Furthermore, cells can sort based upon differences in adhesion between the cells, so even two populations of cells with different levels of the same adhesion molecule can sort out. In cell culture cells that have the strongest adhesion move to the center of a mixed aggregates of cells. Moreover, cell-cell adhesion is often modulated by cell contractility, which can exert forces on the cell-cell contacts so that two cell populations with equal levels of the same adhesion molecule can sort out. The molecules responsible for adhesion are called cell adhesion molecules (CAMs). Several types of cell adhesion molecules are known and one major class of these molecules are cadherins. There are dozens of different cadherins that are expressed on different cell types. Cadherins bind to other cadherins in a like-to-like manner: E-cadherin (found on many epithelial cells) binds preferentially to other E-cadherin molecules. Mesenchymal cells usually express other cadherin types such as N-cadherin.

Extracellular Matrix

The extracellular matrix (ECM) is involved in keeping tissues separated, providing structural support or providing a structure for cells to migrate on. Collagen, laminin, and fibronectin are major ECM molecules that are secreted and assembled into sheets, fibers, and gels. Multisubunit transmembrane receptors called integrins are used to bind to the ECM. Integrins bind extracellularly to fibronectin, laminin, or other ECM components, and intracellularly to microfilament-binding proteins α-actinin and talin to link the cytoskeleton with the outside. Integrins also serve as receptors to trigger signal transduction cascades when binding to the ECM. A well-studied example of morphogenesis that involves ECM is mammary gland ductal branching.

Cell Contractility

Tissues can change their shape and separate into distinct layers via cell contractility. Just as in muscle cells, myosin can contract different parts of the cytoplasm to change its shape or structure. Myosin-driven contractility in embryonic tissue morphogenesis is seen during the separation of germ layers in the model organisms *Caenorhabditis elegans*, *Drosophila* and zebrafish. There are often periodic pulses of contraction in embryonic morphogenesis. A model called the cell state splitter involves alternating cell contraction and expansion, initiated by a bistable organelle at the apical end of each cell. The organelle consists of microtubules and microfilaments in mechanical opposition. It responds to local mechanical perturbations caused by morphogenetic movements. These then trigger traveling embryonic differentiation waves of contraction or expansion over presumptive tissues that determine cell type and is followed by cell differentiation. The cell state splitter was first proposed to explain neural plate morphogenesis during gastrulation of the axolotl and the model was later generalized to all of morphogenesis.

Ecological Genetics

Ecological genetics is the study of the evolution of ecologically relevant traits in natural populations. Ecologically relevant traits are those life history, behavioural, physiological and morphological traits that are related to an individual's fitness or its contribution of offspring to future generations. The term 'ecological genetics' was coined by the English biologist E. B. Ford to describe pioneering research on genetic variation in natural populations that combined field observations of natural selection and adaptive trait variation, experimental evolution and laboratory genetics. Early ecological geneticists like Ford worked to overturn the notion, which prevailed at the time, that natural selection was normally too weak to generate substantial adaptive changes in natural populations. Indeed, Ford concluded that research in ecological genetics had firmly established that 'unexpectedly great selective forces are normally operating to maintain or to adjust the adaptations of organisms in natural conditions'. Today, ecological geneticists continue to advance our understanding of adaptive evolution in natural populations and also apply emerging techniques to reveal the genes underlying important ecological traits and investigate how genetic variation and evolutionary processes shape ecological patterns and processes .Hence, ecological genetics is making advances on two fronts by explaining how species evolve in natural communities and how evolution affects ecological properties such as population abundances and community compositions.

Ecological genetics is nearly synonymous with the field of evolutionary ecology (although the latter is a bit broader) and includes the parts of other subdisciplines such as quantitative and population genetics that apply to adaptive evolution in nature. In general, ecological genetics takes a whole-organism approach to understanding phenotypic evolution, frequently in nonmodel organisms (wild organisms that are difficult to manipulate), and in the process ignores some of the details of the specific changes to deoxyribonucleic acid (DNA) sequences that underlie evolution. This approach can be contrasted with molecular genetics, which focuses on understanding how specific biochemical pathways evolve at the level of nucleic acid substitutions but ignores the details of multitrait phenotypic integration, ecological performance and selection heterogeneity in natural systems. Although the two approaches make simplifying assumptions about evolution, research in ecological and molecular genetics is increasingly integrated, especially as techniques become available to reveal the underlying genetic basis of ecologically relevant traits. The hope is that this integration of molecular and ecological genetics will facilitate progress toward the overarching goal of understanding biology from genes to ecology and back again.

Genetic Materials

Nucleic Acid

Nucleic acid is a naturally occurring chemical compound that is capable of being broken down to yield phosphoric acid, sugars, and a mixture of organic bases (purines and pyrimidines). Nucleic acids are the main information-carrying molecules of the cell, and, by directing the process of protein synthesis, they determine the inherited characteristics of every living thing. The two main classes of nucleic acids are deoxyribonucleic acid (DNA) and ribonucleic acid (RNA). DNA is the master blueprint for life and constitutes the genetic material in all free-living organisms and most viruses. RNA is the genetic material of certain viruses, but it is also found in all living cells, where it plays an important role in certain processes such as the making of proteins.

Nucleotides: Building Blocks of Nucleic Acids

Basic Structure

Nucleic acids are polynucleotides—that is, long chainlike molecules composed of a series of nearly identical building blocks called nucleotides. Each nucleotide consists of a nitrogen-containing aromatic base attached to a pentose (five-carbon) sugar, which is in turn attached to a phosphate group. Each nucleic acid contains four of five possible nitrogen-containing bases: adenine (A), guanine (G), cytosine (C), thymine (T), and uracil (U). A and G are categorized as purines, and C, T, and U are collectively called pyrimidines. All nucleic acids contain the bases A, C, and G; T, however, is found only in DNA, while U is found in RNA. The pentose sugar in DNA (2′-deoxyribose) differs from the sugar in RNA (ribose) by the absence of a hydroxyl group (−OH) on the 2′ carbon of the sugar ring. Without an attached phosphate group, the sugar attached to one of the bases is known as a nucleoside. The phosphate group connects successive sugar residues by bridging the 5′-hydroxyl group on one sugar to the 3′-hydroxyl group of the next sugar in the chain. These nucleoside linkages are called phosphodiester bonds and are the same in RNA and DNA.

Biosynthesis and Degradation

Nucleotides are synthesized from readily available precursors in the cell. The ribose phosphate portion of both purine and pyrimidine nucleotides is synthesized from glucose via the pentose phosphate pathway. The six-atom pyrimidine ring is synthesized first and subsequently attached to the ribose phosphate. The two rings in purines are synthesized while attached to the ribose phosphate during the assembly of adenine or guanine nucleosides. In both cases the end product is a nucleotide carrying a phosphate attached to the 5′ carbon on the sugar. Finally, a specialized enzyme called a kinase adds two phosphate groups using adenosine triphosphate (ATP) as the phosphate donor to form ribonucleoside triphosphate, the immediate precursor of RNA. For DNA, the 2′-hydroxyl group is removed from the ribonucleoside diphosphate to give deoxyribonucleoside diphosphate. An additional phosphate group from ATP is then added by another kinase to form a deoxyribonucleoside triphosphate, the immediate precursor of DNA.

During normal cell metabolism, RNA is constantly being made and broken down. The purine and pyrimidine residues are reused by several salvage pathways to make more genetic material. Purine is salvaged in the form of the corresponding nucleotide, whereas pyrimidine is salvaged as the nucleoside.

Deoxyribonucleic Acid (DNA)

DNA is a polymer of the four nucleotides A, C, G, and T, which are joined through a backbone of alternating phosphate and deoxyribose sugar residues. These nitrogen-containing bases occur in complementary pairs as determined by their ability to form hydrogen bonds between them. A always pairs with T through two hydrogen bonds, and G always pairs with C through three hydrogen bonds. The spans of A:T and G:C hydrogen-bonded pairs are nearly identical, allowing them to bridge the sugar-phosphate chains uniformly. This structure, along with the molecule's chemical stability, makes DNA the ideal genetic material. The bonding between complementary bases also provides a mechanism for the replication of DNA and the transmission of genetic information.

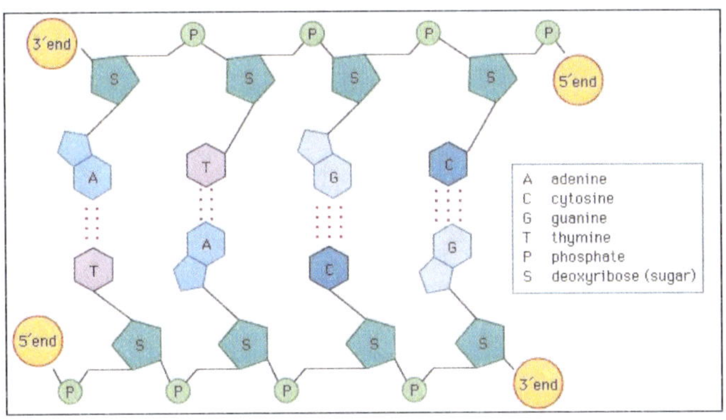

DNA structure

Showing the nucleotide bases cytosine (C), thymine (T), adenine (A), and guanine (G) linked to a backbone of alternating phosphate (P) and deoxyribose sugar (S) groups. Two sugar-phosphate chains are paired through hydrogen bonds between A and T and between G and C, thus forming the twin-stranded double helix of the DNA molecule.

Biochemical Properties

Denaturation

The strands of the DNA double helix are held together by hydrogen bonding interactions between the complementary base pairs. Heating DNA in solution easily breaks these hydrogen bonds, allowing the two strands to separate—a process called denaturation or melting. The two strands may reassociate when the solution cools, reforming the starting DNA duplex—a process called renaturation or hybridization. These processes form the basis of many important techniques for manipulating DNA. For example, a short piece of DNA called an oligonucleotide can be used to test whether a very long DNA sequence has the complementary sequence of the oligonucleotide embedded within it. Using hybridization, a single-stranded DNA molecule can capture complementary sequences from any source. Single strands from RNA can also reassociate. DNA and RNA single strands can form hybrid molecules that are even more stable than double-stranded DNA. These molecules form the basis of a technique that is used to purify and characterize messenger RNA (mRNA) molecules corresponding to single genes.

Ultraviolet Absorption

DNA melting and reassociation can be monitored by measuring the absorption of ultraviolet (UV) light at a wavelength of 260 nanometres (billionths of a metre). When DNA is in a double-stranded conformation, absorption is fairly weak, but when DNA is single-stranded, the unstacking of the bases leads to an enhancement of absorption called hyperchromicity. Therefore, the extent to which DNA is single-stranded or double-stranded can be determined by monitoring UV absorption.

Chemical Modification

After a DNA molecule has been assembled, it may be chemically modified—sometimes deliberately by special enzymes called DNA methyltransferases and sometimes accidentally by oxidation, ionizing radiation, or the action of chemical carcinogens. DNA can also be cleaved and degraded by enzymes called nucleases.

Methylation

Three types of natural methylation have been reported in DNA. Cytosine can be modified either on the ring to form 5-methylcytosine or on the exocyclic amino group to form N^4-methylcytosine. Adenine may be modified to form N^6-methyladenine. N^4-methylcytosine and N^6-methyladenine are found only in bacteria and archaea, whereas 5-methylcytosine is widely distributed. Special enzymes called DNA methyltransferases are responsible for this methylation; they recognize specific sequences within the DNA molecule so that only a subset of the bases is modified. Other methylations of the bases or of the deoxyribose are sometimes induced by carcinogens. These usually lead to mispairing of the bases during replication and have to be removed if they are not to become mutagenic.

Natural methylation has many cellular functions. In bacteria and archaea, methylation forms an essential part of the immune system by protecting DNA molecules from fragmentation by restriction endonucleases. In some organisms, methylation helps to eliminate incorrect base sequences introduced during DNA replication. By marking the parental strand with a methyl group, a cellular mechanism known as the mismatch repair system distinguishes between the newly replicated strand where the errors occur and the correct sequence on the template strand. In higher eukaryotes, 5-methylcytosine controls many cellular phenomena by preventing DNA transcription. Methylation is also believed to signal imprinting, a process whereby some genes inherited from one parent are selectively inactivated. Correct methylation may also repress or activate key genes that control embryonic development. On the other hand, 5-methylcytosine is potentially mutagenic because thymine produced during the methylation process converts C:G pairs to T:A pairs. In mammals, methylation takes place selectively within the dinucleotide sequence CG—a rare sequence, presumably because it has been lost by mutation. In many cancers, mutations are found in key genes at CG dinucleotides.

Nucleases

Nucleases are enzymes that hydrolytically cleave the phosphodiester backbone of DNA. Endonucleases cleave in the middle of chains, while exonucleases operate selectively by degrading from the end of the chain. Nucleases that act on both single- and double-stranded DNA are known.

Restriction endonucleases are a special class that recognize and cleave specific sequences in DNA. Type II restriction endonucleases always cleave at or near their recognition sites. They produce small, well-defined fragments of DNA that help to characterize genes and genomes and that produce recombinant DNAs. Fragments of DNA produced by restriction endonucleases can be moved from one organism to another. In this way it has been possible to express proteins such as human insulin in bacteria.

Mutation

Chemical modification of DNA can lead to mutations in the genetic material. Anions such as bisulfite can deaminate cytosine to form uracil, changing the genetic message by causing C-to-T transitions. Exposure to acid causes the loss of purine residues, though specific enzymes exist in cells to repair these lesions. Exposure to UV light can cause adjacent pyrimidines to dimerize, while oxidative damage from free radicals or strong oxidizing agents can cause a variety of lesions that are mutagenic if not repaired. Halogens such as chlorine and bromine react directly with uracil, adenine, and guanine, giving substituted bases that are often mutagenic. Similarly, nitrous acid reacts with primary amine groups—for example, converting adenosine into inosine—which then leads to changes in base pairing and mutation. Many chemical mutagens, such as chlorinated hydrocarbons and nitrites, owe their toxicity to the production of halides and nitrous acid during their metabolism in the body.

Supercoiling

Circular DNA molecules such as those found in plasmids or bacterial chromosomes can adopt many different topologies. One is active supercoiling, which involves the cleavage of one DNA strand, its winding one or more turns around the complementary strand, and then the resealing of the molecule. Each complete rotation leads to the introduction of one supercoiled turn in the DNA, a process that can continue until the DNA is fully wound and collapses on itself in a tight ball. Reversal is also possible. Special enzymes called gyrases and topoisomerases catalyze the winding and relaxation of supercoiled DNA. In the linear chromosomes of eukaryotes, the DNA is usually tightly constrained at various points by proteins, allowing the intervening stretches to be supercoiled. This property is partially responsible for the great compaction of DNA that is necessary to fit it within the confines of the cell. The DNA in one human cell would have an extended length of between two and three metres, but it is packed very tightly so that it can fit within a human cell nucleus that is 10 micrometres in diameter.

Sequence Determination

Methods to determine the sequences of bases in DNA were pioneered in the 1970s by Frederick Sanger and Walter Gilbert, whose efforts won them a Nobel Prize in 1980. The Gilbert-Maxam method relies on the different chemical reactivities of the bases, while the Sanger method is based on enzymatic synthesis of DNA in vitro. Both methods measure the distance from a fixed point on DNA to each occurrence of a particular base—A, C, G, or T. DNA fragments obtained from a series of reactions are separated according to length in four "lanes" by gel electrophoresis. Each lane corresponds to a unique base, and the sequence is read directly from the gel. The Sanger method has now been automated using fluorescent dyes to label the DNA, and a single machine can produce tens of thousands of DNA base sequences in a single run.

Ribonucleic Acid (RNA)

RNA is a single-stranded nucleic acid polymer of the four nucleotides A, C, G, and U joined through a backbone of alternating phosphate and ribose sugar residues. It is the first intermediate in converting the information from DNA into proteins essential for the working of a cell. Some RNAs also serve direct roles in cellular metabolism. RNA is made by copying the base sequence of a section of double-stranded DNA, called a gene, into a piece of single-stranded nucleic acid. This process, called transcription, is catalyzed by an enzyme called RNA polymerase.

Types of RNA

Messenger RNA (mRNA)

Messenger RNA (mRNA) delivers the information encoded in one or more genes from the DNA to the ribosome, a specialized structure, or organelle, where that information is decoded into a protein. In prokaryotes, mRNAs contain an exact transcribed copy of the original DNA sequence with a terminal 5′-triphosphate group and a 3′-hydroxyl residue. In eukaryotes the mRNA molecules are more elaborate. The 5′-triphosphate residue is further esterified, forming a structure called a cap. At the 3′ ends, eukaryotic mRNAs typically contain long runs of adenosine residues (polyA) that are not encoded in the DNA but are added enzymatically after transcription. Eukaryotic mRNA molecules are usually composed of small segments of the original gene and are generated by a process of cleavage and rejoining from an original precursor RNA (pre-mRNA) molecule, which is an exact copy of the gene. In general, prokaryotic mRNAs are degraded very rapidly, whereas the cap structure and the polyA tail of eukaryotic mRNAs greatly enhance their stability.

Ribosomal RNA (rRNA)

Ribosomal RNA (rRNA) molecules are the structural components of the ribosome. The rRNAs form extensive secondary structures and play an active role in recognizing conserved portions of mRNAs and tRNAs. They also assist with the catalysis of protein synthesis. In the prokaryote E. coli, seven copies of the rRNA genes synthesize about 15,000 ribosomes per cell. In eukaryotes the numbers are much larger. Anywhere from 50 to 5,000 sets of rRNA genes and as many as 10 million ribosomes may be present in a single cell. In eukaryotes these rRNA genes are looped out of the main chromosomal fibres and coalesce in the presence of proteins to form an organelle called the nucleolus. The nucleolus is where the rRNA genes are transcribed and the early assembly of ribosomes takes place.

Transfer RNA (tRNA)

Transfer RNA (tRNA) carries individual amino acids into the ribosome for assembly into the growing polypeptide chain. The tRNA molecules contain 70 to 80 nucleotides and fold into a characteristic cloverleaf structure. Specialized tRNAs exist for each of the 20 amino acids needed for protein synthesis, and in many cases more than one tRNA for each amino acid is present. The nucleotide sequence is converted into a protein sequence by translating each three-base sequence (called a codon) with a specific protein. The 61 codons used to code amino acids can be read by many fewer than 61 distinct tRNAs. In *E. coli* a total of 40 different tRNAs are used to translate the 61 codons. The amino acids are loaded onto the tRNAs by specialized enzymes called aminoacyl tRNA synthetases, usually with one synthetase for each amino acid. However, in some organisms, less than the full complement of 20 synthetases are required because some amino acids, such as glutamine and asparagine, can be synthesized on their respective tRNAs. All tRNAs adopt similar structures because they all have to interact with the same sites on the ribosome.

Ribozymes

Not all catalysis within the cell is carried out exclusively by proteins. Thomas Cech and Sidney Altman, jointly awarded a Nobel Prize in 1989, discovered that certain RNAs, now known as ribozymes, showed enzymatic activity. Cech showed that a noncoding sequence (intron) in the small subunit rRNA of protozoans, which had to be removed before the rRNA was functional, can excise itself from a much longer precursor RNA molecule and rejoin the two ends in an autocatalytic reaction. Altman showed that the RNA component of an RNA protein complex called ribonuclease P can cleave a precursor tRNA to generate a mature tRNA. In addition to self-splicing RNAs similar to the one discovered by Cech, artificial RNAs have been made that show a variety of catalytic reactions. It is now widely held that there was a stage during evolution when only RNA catalyzed and stored genetic information. This period, sometimes called "the RNA world," is believed to have preceded the function of DNA as genetic material.

Antisense RNAs

Most antisense RNAs are synthetically modified derivatives of RNA or DNA with potential therapeutic value. In nature, antisense RNAs contain sequences that are the complement of the normal coding sequences found in mRNAs (also called sense RNAs). Like mRNAs, antisense RNAs are single-stranded, but they cannot be translated into protein. They can inactivate their complementary mRNA by forming a double-stranded structure that blocks the translation of the base sequence. Artificially introducing antisense RNAs into cells selectively inactivates genes by interfering with normal RNA metabolism.

Viral Genomes

Many viruses use RNA for their genetic material. This is most prevalent among eukaryotic viruses, but a few prokaryotic RNA viruses are also known. Some common examples include poliovirus, human immunodeficiency virus (HIV), and influenza virus, all of which affect humans, and tobacco mosaic virus, which infects plants. In some viruses the entire genetic material is encoded in a single RNA molecule, while in the segmented RNA viruses several RNA molecules may be present. Many RNA viruses such as HIV use a specialized enzyme called reverse transcriptase that permits replication of the virus through a DNA intermediate. In some cases this DNA intermediate becomes integrated into the host chromosome during infection; the virus then exists in a dormant state and effectively evades the host immune system.

Other RNAs

Many other small RNA molecules with specialized functions are present in cells. For example, small nuclear RNAs (snRNAs) are involved in RNA splicing, and other small RNAs that form part of the enzymes telomerase or ribonuclease P are part of ribonucleoprotein particles. The RNA component of telomerase contains a short sequence that serves as a template for the addition of small strings of oligonucleotides at the ends of eukaryotic chromosomes. Other RNA molecules serve as guide RNAs for editing, or they are complementary to small sections of rRNA and either direct the positions at which methyl groups need to be added or mark U residues for conversion to the isomer pseudouridine.

Genome

A genome is the complete set of genetic information in an organism. It provides all of the information the organism requires to function. In living organisms, the genome is stored in long molecules of DNA called chromosomes. Small sections of DNA, called genes, code for the RNA and protein molecules required by the organism. In eukaryotes, each cell's genome is contained within a membrane-bound structure called the nucleus. Prokaryotes, which contain no inner membranes, store their genome in a region of the cytoplasm called the nucleoid. The full range of RNA molecules expressed by a genome is known as its transcriptome, and the full assortment of proteins produced by the genome is called its proteome.

There are 23 pairs of chromosomes in the human genome. Between 1990 and 2003, all twenty-three pairs were fully sequenced through an international research undertaking known as the Human Genome Project. The study and analysis of genomes is called genomics.

Genetic Code

The genetic code is the code our body uses to convert the instructions contained in our DNA the essential materials of life. It is typically discussed using the "codons" found in mRNA, as mRNA is the messenger that carries information from the DNA to the site of protein synthesis.

Everything in our cells is ultimately built based on the genetic code. Our hereditary information – that is, the information that's passed down from parent to child – is stored in the form of DNA. That DNA is then used to build RNA, proteins, and ultimately cells, tissues, and organs.

Like binary code, DNA uses a chemical language with just a few letters to store information in a very efficient manner. While binary uses only ones and zeroes, DNA has four letters – the four nucleotides Adenine, Cytosine, Guanine, and Thymine/Uracil.

Thymine and Uracil are very similar to each other, except that "Thymine" is slightly more stable and is used in DNA. Uracil is used in RNA, and has all the same properties of Thymine except that it is slightly more prone to mutate.

This doesn't matter in RNA, since new RNA copies can be produced from DNA at any time, and most RNA molecules are intentionally destroyed by the cell a short time after they're produced so that the cell does not waste resources producing unneeded proteins from old RNA molecules.

Together, these four letters of A, C, G, and T/U are used to "spell" coded instructions for each amino acid, as well as other instructions like "start transcription" and "stop transcription."

Instructions for "start," "stop," or for a given amino acid are "read" by the cell in three-letter blocks called "codons." When we talk about "codons," we usually mean codons in mRNA – the "messenger RNA" that is made by copying the information in DNA.

For that reason, we talk about codons made of RNA, which uses Uracil, instead of the original DNA code which uses Thymine.

One of the most remarkable evidences for the common descent of all life on Earth from a single ancestor is the fact that all organisms use the same genetic code to translate DNA into amino acids.

There are a few slight exceptions to be found, but the genetic code is similar enough across organisms that when a gene from a plant or jellyfish is injected into a mammal cell, for example, the mammal cell will read the gene in the same way and build the same product as the original plant or jellyfish.

Function of Genetic Code

The genetic code allows cells to contain a mind-boggling amount of information.

A microscopic fertilized egg cell, following the instructions contained in its genetic code, can produce a human or elephant which even has similar personality and behaviors to those of its parents. There is a lot of information in there.

The development of the genetic code was vital because it allowed living things to reliably produce products necessary for their survival – and pass instructions for how to do the same onto the next generation.

When a cell seeks to reproduce, one of the first things it does is make a copy of its DNA. This is the "S" phase of the cell cycle, which stands for the "Synthesis" of a new copy of the cell's DNA.

The information encoded in DNA is preserved by the specific pairing of DNA bases with each other. Adenine will only bond with Thymine, Cytosine with Guanine, etc.

That means that when a cell wants to copy its DNA, all it has to do is part the two strands of the double helix and line up the nucleotides that the bases of the existing DNA "want" to pair with.

This specific base pairing ensures that the new partner strand will contain the same sequence of base pairs – the same "code" – as the old partner strand. Each resulting double helix contains one strand of old DNA paired with one strand of new DNA.

These new double helixes will be inherited by two daughter cells. When it's time for those daughter cells to reproduce, each strand of these new double helices, act as templates for a new double helix.

When the time comes for a cell to "read" the instructions contained within its DNA, it uses the same principle of specific pair bonding. RNA is very similar to DNA, and each RNA base bonds specifically to one DNA base. Uracil binds to Adenine, Cytosine to Guanine, etc.

This means that, just like DNA replication, the information in DNA is accurately transferred to RNA as long as the resulting RNA strand is composed of the bases that bind specifically with the bases in the DNA.

Sometimes, the RNA strand itself can be the end product. Structures made of RNA perform important functions in ourselves, including assembling proteins, regulating gene expression, and catalyzing the formation of proteins.

In fact, some scientists think that the first life on Earth might have been composed mainly of RNA. This is because RNA can store information in its base pairs just like DNA, but can also perform some enzymatic and regulatory functions.

In most cases, however, the RNA goes on to be transcribed into a protein. Using the amino acid "building blocks of life," our cells can build almost protein machines for almost any purpose, from muscle fibers to neurotransmitters to digestive enzymes.

In protein transcription, the RNA codons that were transcribed from the DNA are "read" by a ribosome. The ribosome finds the appropriate transfer RNA (tRNA) with "anti-codons" that are complimentary to the codons in the messenger RNA (mRNA) that has been transcribed from the DNA.

Ribosomes catalyze the formation of peptide bonds between the amino acids as they "read" each codon in the mRNA. At the end of the process, you have a string of amino acids as specified by the DNA – that is, a protein.

Other building blocks of life, such as sugars and lipids, are in turn created by proteins. In this way the information contained in the DNA is transformed into all of the materials of life, using the genetic code.

Genome Mapping

Genome mapping is used to identify and record the location of genes and the distances between genes on a chromosome. Genome mapping provided a critical starting point for the Human Genome Project.

- A genome map highlights the key 'landmarks' in an organism's genome.
- A bit like how the London tube map shows the different stops on a tube line to

help you get around the city, a genome map helps scientists to navigate their way around the genome.

- The landmarks on a genome map may include short DNA sequences, regulatory sites that turn genes on and off or the genes themselves.
- Genome mapping provided the basis for whole genome sequencing and the Human Genome Project.
- Sequenced DNA fragments can be aligned to the genome map to aid with the assembly of the genome.
- Over time, as scientists learn more about a particular genome, its map becomes more accurate and detailed. A genome map is not a final product, but work in progress.

Different Types of Genome Mapping

- There are two general types of genome mapping called genetic mapping and physical mapping.
- Both types of genome mapping guide scientists towards the location of a gene (or section of DNA) on a chromosome, however, they rely on very different information.
    - Genetic mapping looks at how genetic information is shuffled between chromosomes or between different regions in the same chromosome during meiosis (a type of cell division). A process called recombination or 'crossing over'.
    - Physical mapping looks at the physical distance between known DNA sequences (including genes) by working out the number of base pairs (A-T, C-G) between them.

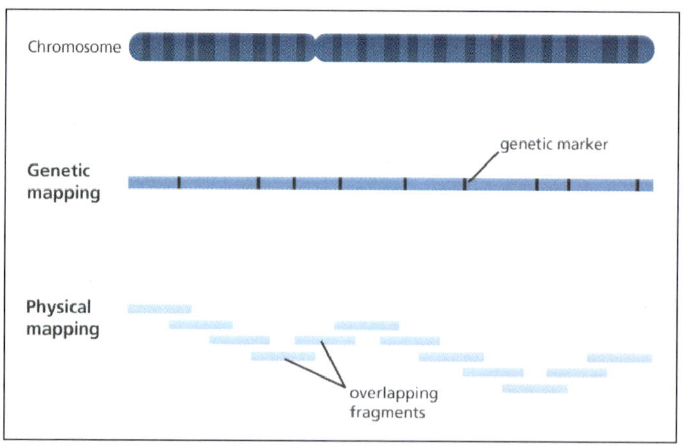

Illustration showing the difference between the two basic ways of mapping a genome: genetic mapping and physical mapping.

Genetic Mapping

Early Genetic Maps

- Alfred Sturtevant created the first genetic map of a chromosome from the fruit fly (Drosophila melanogaster) in 1913.
  - He determined that genes were arranged on chromosomes in a linear way, like beads on a necklace, and that genes for specific traits are located in particular places.
  - He proposed that the frequency of 'crossing over' (recombination) between two genes could help determine their location on a chromosome.
  - He realised that genes that were far apart on a chromosome are more likely to be inherited separately simply because there is a larger region over which recombination can occur.
  - In the same way, genes that are close to each other on the chromosome are more likely to be inherited together.

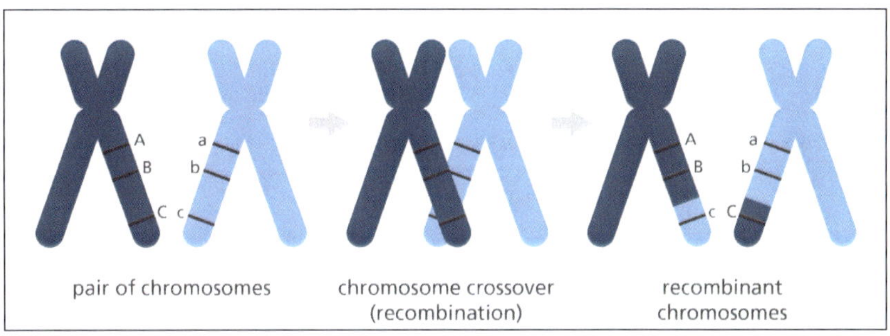

Illustration showing crossing over of chromosomes during meiosis and how this affects the likelihood of genes being inherited together.

- By finding out how often various characteristics are inherited together it is possible to estimate the distance between the genes. A map of where the genes are in relationship to each other on the chromosomes can then be drawn. This is called a linkage map.
- Genes that are on the same chromosome are said to be 'linked' and the distance between these genes is called a 'linkage distance'. The smaller the distance the more likely two genes will be inherited together.
- The same concept of studying how traits are passed on was applied to develop the first human genome map.
- If two (or more) characteristics were seen to be frequently inherited together in a family, for example blonde hair and blue eyes, it suggested that the genes for the two characteristics were close together on a particular chromosome.

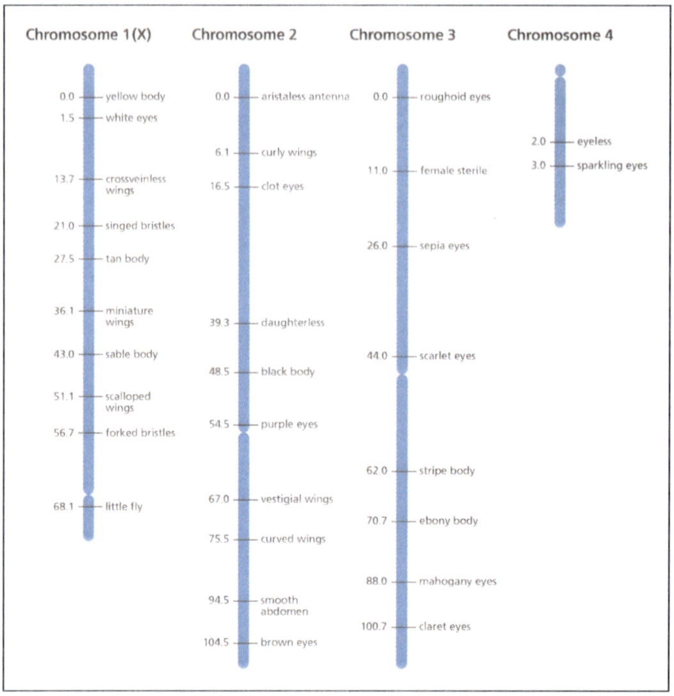

Illustration showing a genetic map of the chromosomes from the fruit fly (Drosophila melanogaster). The names of the genes are shown to the right of each chromosome. The numbers to the left of each chromosome represent the distance between these genes.

Modern Genetic Maps

- With more recent genetic mapping techniques, the position of genes is worked out from finding the exact frequency of genetic recombination that has occurred.
- To produce a genetic map, researchers collect blood or tissue samples from members of a family, some of whom have a certain disease or characteristic.
- The researchers then isolate the DNA from samples taken from each individual and closely examine it to find unique patterns in the DNA of those individuals with the disease/characteristic, that aren't present in the DNA of the individuals who don't have the disease/characteristic.
- These are referred to as markers and are extremely valuable for tracking inheritance of characteristics or diseases through several generations of a family.
- One type of DNA marker, called a microsatellite, is found throughout the genome and consists of a specific repeated sequence of bases.
- If a particular gene is close to a DNA marker on the chromosome, it is more likely that the gene and marker will stay together during the recombination process

and are therefore more likely to be passed down along the family line (inherited) together.

- In the same way, if a DNA marker and gene are frequently separated by the recombination process it suggests that they are far apart on the chromosome and are less likely to be inherited together.
- The more DNA markers there are on a genetic map the more likely it is that one of them will be located close to the disease or trait-associated gene.
- While genetic maps are good at giving you the bigger picture, they have limited accuracy and therefore need to be supplemented with further information gained from other mapping techniques, such as physical mapping.

Physical Mapping

- Physical mapping gives an estimation of the (physical) distance between specific known DNA sequences on a chromosome.
- The distance between these known DNA sequences on a chromosome is expressed as the number of base pairs between them.
- There are a several different techniques used for physical mapping. These include:
    - Restriction mapping (fingerprint mapping and optical mapping).
    - Fluorescent in situ hybridisation (FISH) mapping.
    - Sequence tagged site (STS) mapping.

Restriction Mapping

- This uses specific restriction enzymes? to cut an unknown segment of DNA at short, known base sequences called restriction sites.
- Restriction enzymes always cut DNA at a specific sequence of DNA (restriction site). For example, the restriction enzyme EcoRI (taken from Escherichia coli) always cuts at the sequence GAATTC/CTTAAG. Therefore if we use EcoRI to cut the DNA we know that the DNA sequence either side of the cut will be AATT.
- A restriction map shows all the locations of that particular restriction site (GAATTC) throughout the genome.
- A physical map is generated by aligning the different restriction maps along the chromosomes.
- There a two specific types of restriction mapping – optical and fingerprint.

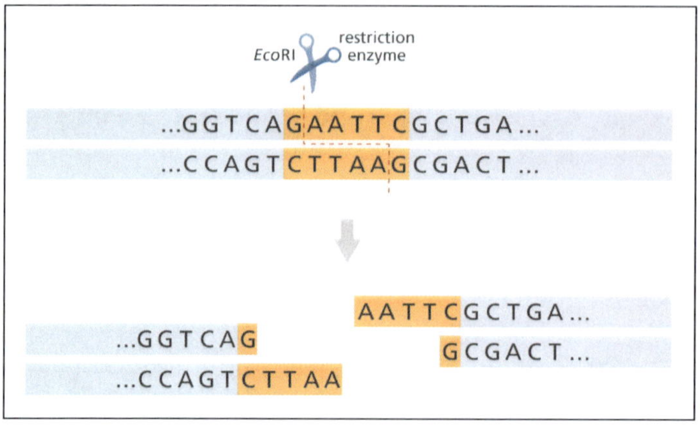

Illustration showing the restriction site for the restriction enzyme EcoRI. Restriction enzymes always cut DNA at a specific sequence of DNA.

Fingerprint Mapping

- In fingerprint mapping the genome is broken into fragments which are then copied in bacteria and cells.
- The DNA copies (clones) are then cut by restriction enzymes and the lengths of the resulting fragments are estimated using a lab method called electrophoresis.
- Electrophoresis separates the fragments of DNA according to size resulting in a distinct banding pattern.

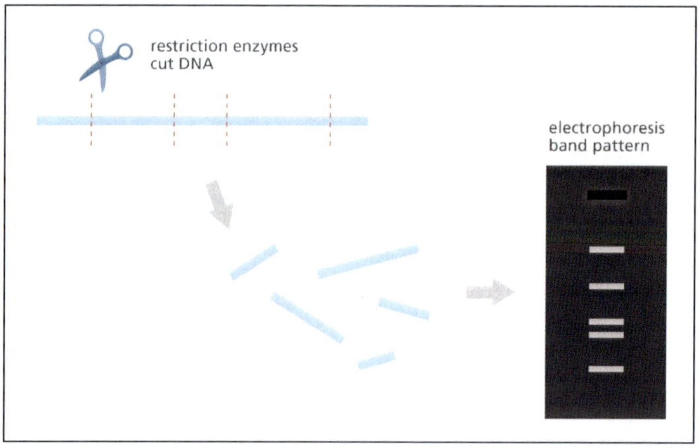

Illustration showing how a DNA fingerprint is created by electrophoresis.

- The fingerprint map is constructed by comparing the patterns from all the fragments of DNA to find areas of similarity. Those with similar patterns are then grouped together to form a map.
- Fingerprint mapping formed the basis to the sequencing of the human, mouse, zebrafish and pig genomes.

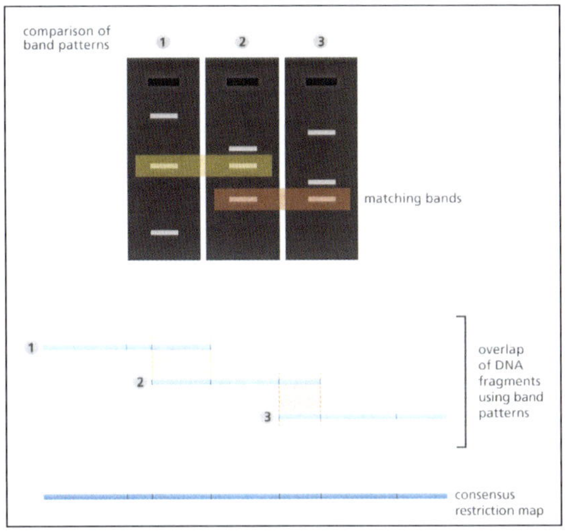

Illustration showing how DNA fingerprints can be compared to produce a genome map.

Optical Mapping

- Optical mapping uses single molecules of DNA that are stretched and held in place on a slide.
- Restriction enzymes are added to cut the DNA at specific points leaving gaps behind.
- The fragments are then stained with dye and the gaps are visualised under a fluorescence microscope.
- The intensity of the fluorescence is used to construct an optical map of single molecules.
- These can then be combined and overlapped to give a global overview of the genome and aid with assembling a sequenced genome.

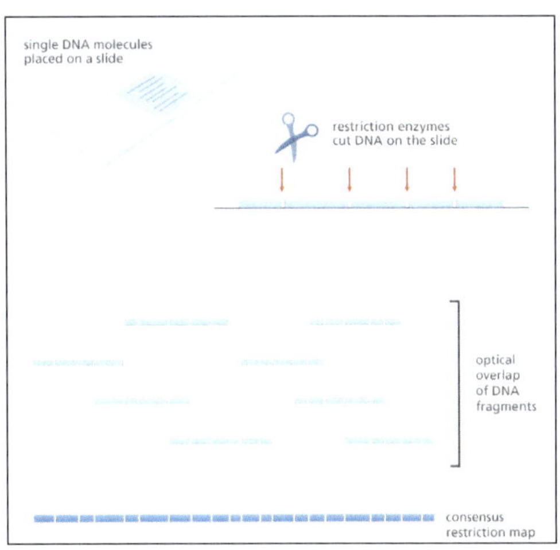

Illustration showing the process of optical mapping.

Fluorescent in Situ Hybridisation (FISH) Mapping

- This uses fluorescent probes to detect the location of DNA sequences on chromosomes.
- First, the probes are prepared. These are short sequences of single-stranded DNA, that match the DNA sequence that the scientist wants to find.
- The probes are then labelled with fluorescent dye before being mixed with the chromosome DNA so that it can bind to a complementary strand of DNA on the chromosome.
- The fluorescent tag allows the scientist to see the location of the DNA sequence on the chromosome.

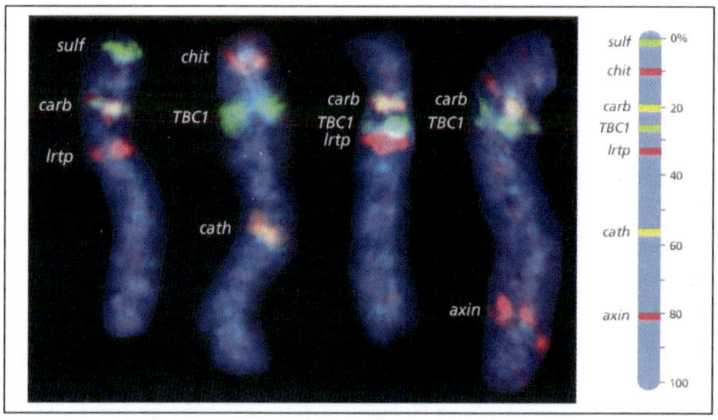

Illustration showing how FISH can be used to produce a genetic map.

The photograph on the left shows Chromosome 17 from four British peppered moths with fluorescent probes indicating the physical positions of specific genes. The illustration on the right shows the relative positions of the genes on the chromosome.

Sequence-tagged Site Mapping

- This technique maps the positions of short DNA sequences (between 200-500 base pairs in length) that are easily recognisable and only occur once in the genome. These short DNA sequences are called sequence-tagged sites (STSs).
- To map a set of STSs a collection of overlapping DNA fragments from a single chromosome or the entire genome is required.
- To do this, the genome is first broken up into fragments.
- The fragments are then replicated up to 10 times in bacterial cells to create a library of DNA clones.
- The polymerase chain reaction (PCR) is then used to determine which frag-

ments contain STSs. Special primers are designed to bind either side of the STS to ensure that only that part of the DNA is copied.

- If two DNA fragments are found to contain the same STS then they must represent overlapping parts of the genome.
- If one DNA fragment contains two different STSs then those two STSs must be near to each other in the genome.

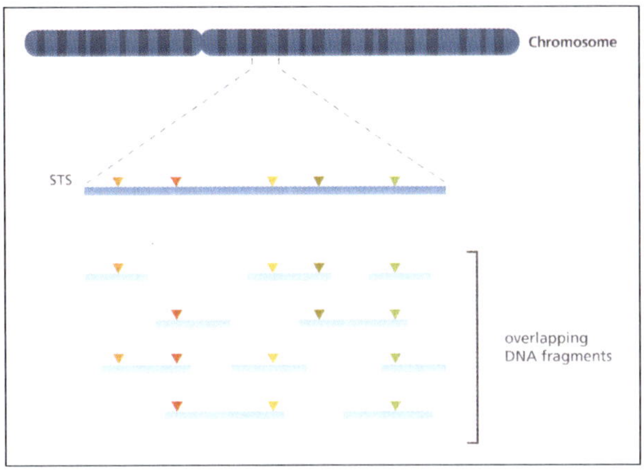

Illustration showing the process of STS mapping.

References

- Genetics, science: britannica.com, Retrieved 2 April, 2019
- Ecogenetics: safeopedia.com, Retrieved 18 January, 2019
- Mortlock, Robert (2013). Microorganisms As Model Systems for Studying Evolution. Springer Verlag. p. 2. ISBN 978-1-4684-4846-7
- Developmental-genetics: le.ac.uk, Retrieved 7 May, 2019
- Cell-division-and-growth, cell-biology, science: britannica.com, Retrieved 13 February, 2019
- Morphogenesis, science: britannica.com, Retrieved 3 July, 2019
- Thompson, D'Arcy Wentworth (1992), On Growth and Form: The Complete Revised Edition, New York, NY: Dover, ISBN 978-0-486-67135-2
- Nucleic-acid, science: britannica.com, Retrieved 20 March, 2019
- Weeks, Benjamin S. (2012). Alcamo's microbes and society (3rd ed.). Sudbury, MA: Jones & Bartlett Learning. ISBN 978-0-7637-9064-6
- Genome: nature.com, Retrieved 25 August, 2019
- Genetic-code: biologydictionary.net, Retrieved 23 June, 2019
- Bernstein, Harris; Bernstei, Carol (2013). "Evolutionary Origin and Adaptive Function of Meiosis". Meiosis. doi:10.5772/56557. ISBN 978-953-51-1197-9
- How-do-you-map-a-genome: yourgenome.org, Retrieved 14 January, 2019

Chapter 2
Heredity and Chromosomes

The passing on of traits from parents to their offspring is known as heredity. It takes place either through sexual reproduction or asexual reproduction. A chromosome is a DNA molecule that consists of a part or all the genetic material of an organism. This chapter discusses in detail the concepts and processes related to heredity and chromosomes such as the behavior of chromosomes during cell division, chromosome mutations and Mendelian inheritance.

Heredity

Heredity is the sum of all biological processes by which particular characteristics are transmitted from parents to their offspring. The concept of heredity encompasses two seemingly paradoxical observations about organisms: the constancy of a species from generation to generation and the variation among individuals within a species. Constancy and variation are actually two sides of the same coin, as becomes clear in the study of genetics. Both aspects of heredity can be explained by genes, the functional units of heritable material that are found within all living cells. Every member of a species has a set of genes specific to that species. It is this set of genes that provides the constancy of the species. Among individuals within a species, however, variations can occur in the form each gene takes, providing the genetic basis for the fact that no two individuals (except identical twins) have exactly the same traits.

The set of genes that an offspring inherits from both parents, a combination of the genetic material of each, is called the organism's genotype. The genotype is contrasted to the phenotype, which is the organism's outward appearance and the developmental outcome of its genes. The phenotype includes an organism's bodily structures, physiological processes, and behaviours. Although the genotype determines the broad limits of the features an organism can develop, the features that actually develop, i.e., the phenotype, depend on complex interactions between genes and their environment. The genotype remains constant throughout an organism's lifetime; however, because the organism's internal and external environments change continuously, so does its phenotype. In conducting genetic studies, it is crucial to discover the degree to which the observable trait is attributable to the pattern of genes in the cells and to what extent it arises from environmental influence.

Because genes are integral to the explanation of hereditary observations, genetics also can be defined as the study of genes. Discoveries into the nature of genes have shown that genes are important determinants of all aspects of an organism's makeup. For this reason, most areas of biological research now have a genetic component, and the study of genetics has a position of central importance in biology. Genetic research also has demonstrated that virtually all organisms on this planet have similar genetic systems, with genes that are built on the same chemical principle and that function according to similar mechanisms. Although species differ in the sets of genes they contain, many similar genes are found across a wide range of species. For example, a large proportion of genes in baker's yeast are also present in humans. This similarity in genetic makeup between organisms that have such disparate phenotypes can be explained by the evolutionary relatedness of virtually all life-forms on Earth. This genetic unity has radically reshaped the understanding of the relationship between humans and all other organisms. Genetics also has had a profound impact on human affairs. Throughout history humans have created or improved many different medicines, foods, and textiles by subjecting plants, animals, and microbes to the ancient techniques of selective breeding and to the modern methods of recombinant DNA technology. In recent years medical researchers have begun to discover the role that genes play in disease. The significance of genetics only promises to become greater as the structure and function of more and more human genes are characterized.

Basic Features of Heredity

Mendelian Genetics

Gregor Mendel published his work in the proceedings of the local society of naturalists in Brünn, Austria (now Brno, Czech Republic), in 1866, but none of his contemporaries appreciated its significance. It was not until 1900, 16 years after Mendel's death, that his work was rediscovered independently by botanists Hugo de Vries in Holland, Carl Erich Correns in Germany, and Erich Tschermak von Seysenegg in Austria. Like several investigators before him, Mendel experimented on hybrids of different varieties of a plant; he focused on the common pea plant (*Pisum sativum*). His methods differed in two essential respects from those of his predecessors. First, instead of trying to describe the appearance of whole plants with all their characteristics, Mendel followed the inheritance of single, easily visible and distinguishable traits, such as round versus wrinkled seed, yellow versus green seed, purple versus white flowers, and so on. Second, he made exact counts of the numbers of plants bearing each trait; it was from such quantitative data that he deduced the rules governing inheritance.

Since pea plants reproduce usually by self-pollination of their flowers, the varieties Mendel obtained from seedsmen were "pure"—i.e., descended for several to many generations from plants with similar traits. Mendel crossed them by deliberately transferring the pollen of one variety to the pistils of another; the resulting first-generation hybrids, denoted by the symbol F_1, usually showed the traits of only one parent. For example, the crossing of yellow-seeded plants with green-seeded ones gave yellow seeds, and the crossing of purple-flowered plants with white-flowered ones gave purple-flowered plants. Traits such as the yellow-seed colour and the purple-flower colour Mendel called dominant; the green-seed colour and the white-flower colour he called recessive. It looked as if the yellow and purple "bloods" overcame or consumed the green and white "bloods."

That this was not so became evident when Mendel allowed the F_1 hybrid plants to self-pollinate and produce the second hybrid generation, F_2. Here, both the dominant and the recessive traits reappeared, as pure and uncontaminated as they were in the original parents (generation P). Moreover, these traits now appeared in constant proportions: about 3/4 of the plants in the second generation showed the dominant trait and 1/4 showed the recessive, a 3 to 1 ratio. It can be seen in the table that Mendel's actual counts were as close to the ideal ratio as one could expect, allowing for the sampling deviations present in all statistical data.

Pea plants with dominant and recessive characters obtained by Mendel in the second generation of hybrids				
Number Dominant		**Number Recessive**		**Ratio**
Round Seed	5,474	Wrinkled Seed	1,850	2.96:1
Yellow Seed	6,022	Green Seed	2,001	3.01:1
Purple Flowers	705	White Flowers	224	3.15:1
Tall Plants	787	Short Plants	277	2.84:1

Mendel concluded that the sex cells, the gametes, of the purple-flowered plants carried some factor that caused the progeny to develop purple flowers, and the gametes of the white-flowered variety had a variant factor that induced the development of white flowers. In 1909 the Danish biologist Wilhelm Ludvig Johannsen proposed to call these factors genes.

An example of one of Mendel's experiments will illustrate how the genes are transmitted and in what particular ratios. Let *R* stand for the gene for purple flowers and *r* for the gene for white flowers (dominant genes are conventionally symbolized by capital letters and recessive genes by lowercase letters). Since each pea plant contains a gene endowment half of whose set is derived from the mother and half from the father, each plant has two genes for flower colour. If the two genes are alike—for instance, both having come from white-flowered parents (*rr*)—the plant is termed a homozygote. The union of gametes with different genes gives a hybrid plant, termed a heterozygote (*Rr*). Since the gene *R*, for purple, is dominant over *r*, for white, the F_1 generation hybrids will show purple flowers. They are phenotypically purple, but their genotype contains both *R* and *r* genes, and these alternative (allelic or allelomorphic) genes do not blend or contaminate each other. Mendel inferred that, when a heterozygote forms its sex cells, the allelic genes segregate and pass to different gametes. This is expressed in the first law of Mendel, the law of segregation of unit genes. Equal numbers of gametes, ovules, or pollen grains are formed that contain the genes *R* and *r*. Now, if the gametes unite at random, then the F_2 generation should contain about $^1/_4$ white-flowered and $^3/_4$ purple-flowered plants. The white-flowered plants, which must be recessive homozygotes, bear the genotype *rr*. About $^1/_3$ of the plants exhibiting the dominant trait of purple flowers must be homozygotes, *RR*, and $^2/_3$ heterozygotes, *Rr*. The prediction is tested by obtaining a third generation, F_3, from the purple-flowered plants; though phenotypically all purple-flowered, $^2/_3$ of this group of plants reveal the presence of the recessive gene allele, *r*, in their genotype by producing about $^1/_4$ white-flowered plants in the F_3 generation.

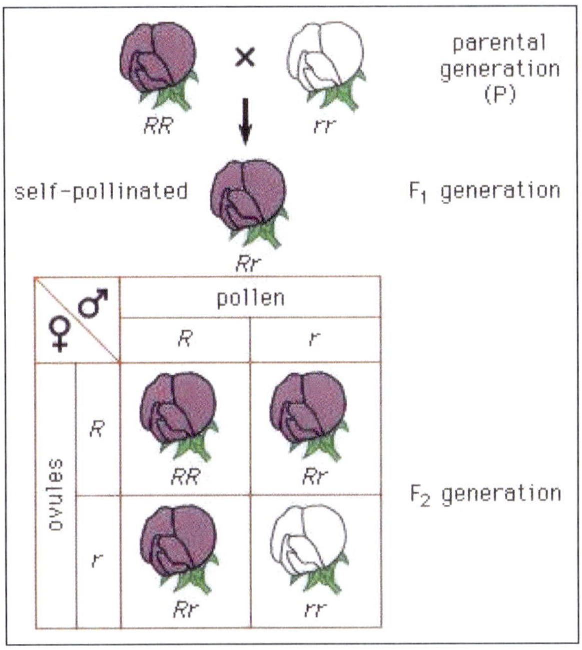

Mendel's law of segregation.

In above figure, cross of a purple-flowered and a white-flowered strain of peas; R stands for the gene for purple flowers and r for the gene for white flowers.

Mendel also crossbred varieties of peas that differed in two or more easily distinguishable traits. When a variety with yellow round seed was crossed to a green wrinkled-seed variety, the F_1 generation hybrids produced yellow round seed. Evidently, yellow (*A*) and round (*B*) are dominant traits, and green (*a*) and wrinkled (*b*) are recessive. By allowing the F_1 plants (genotype *AaBb*) to self-pollinate, Mendel obtained an F_2 generation of 315 yellow round, 101 yellow wrinkled, 108 green round, and 32 green wrinkled seeds, a ratio of approximately 9 : 3 : 3 : 1. The important point here is that the segregation of the colour (*A–a*) is independent of the segregation of the trait of seed surface (*B–b*). This is expected if the F_1 generation produces equal numbers of four kinds of gametes, carrying the four possible combinations of the parental genes: *AB*, *Ab*, *aB*, and *ab*. Random union of these gametes gives, then, the four phenotypes in a ratio 9 dominant–dominant : 3 recessive–dominant : 3 dominant–recessive : 1 recessive–recessive. Among these four phenotypic classes there must be nine different genotypes, a supposition that can be tested experimentally by raising a third hybrid generation. The predicted genotypes are actually found. Another test is by means of a backcross (or testcross); the F_1 hybrid (phenotype yellow round seed; genotype *AaBb*) is crossed to a double recessive plant (phenotype green wrinkled seed; genotype *aabb*). If the hybrid gives four kinds of gametes in equal numbers and if all the gametes of the double recessive are alike (*ab*), the predicted progeny of the backcross are yellow round, yellow wrinkled, green round, and green wrinkled seed in a ratio 1 : 1 : 1 : 1. This prediction is realized in experiments. When the varieties crossed differ in three genes, the F_1 hybrid forms 2^3, or eight, kinds of gametes (2^n = kinds of gametes, *n*being the number of genes). The second generation of hybrids, the F_2, has 27 (3^3) genotypically distinct kinds of individuals but only eight different phenotypes. From these results and others, Mendel derived his second law: the law of recombination, or independent assortment of genes.

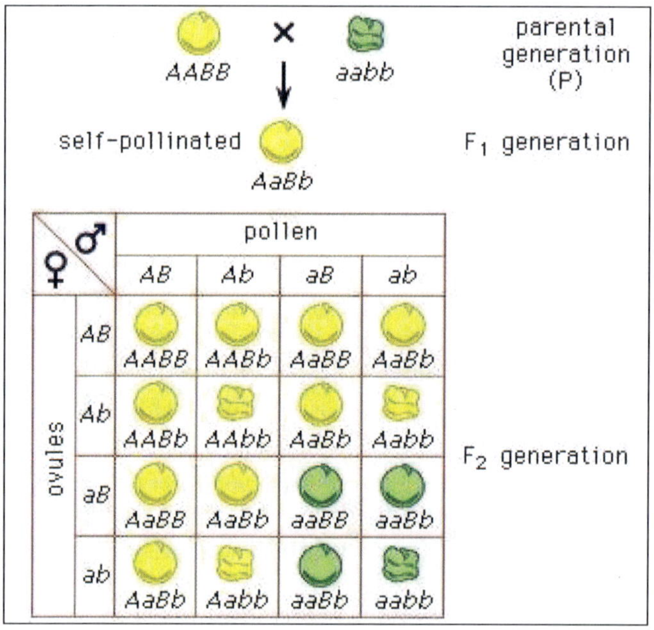

Mendel's law of independent assortment.

In above figure, cross of peas having yellow round seeds with peas having green wrinkled seeds. A stands for the gene for yellow and a for the gene for green; B stands for the gene for a round surface and b for the gene for a wrinkled surface.

Universality of Mendel's laws

Although Mendel experimented with varieties of peas, his laws have been shown to apply to the inheritance of many kinds of characters in almost all organisms. In 1902 Mendelian inheritance was demonstrated in poultry (by English geneticists William Bateson and Reginald Punnett) and in mice. The following year, albinism became the first human trait shown to be a Mendelian recessive, with pigmented skin the corresponding dominant.

In 1902 and 1909, English physician Sir Archibald Garrod initiated the analysis of inborn errors of metabolism in humans in terms of biochemical genetics. Alkaptonuria, inherited as a recessive, is characterized by excretion in the urine of large amounts of the substance called alkapton, or homogentisic acid, which renders the urine black on exposure to air. In normal (i.e., nonalkaptonuric) persons the homogentisic acid is changed to acetoacetic acid, the reaction being facilitated by an enzyme, homogentisic acid oxidase. Garrod advanced the hypothesis that this enzyme is absent or inactive in homozygous carriers of the defective recessive alkaptonuria gene; hence, the homogentisic acid accumulates and is excreted in the urine. Mendelian inheritance of numerous traits in humans has been studied since then.

In analyzing Mendelian inheritance, it should be borne in mind that an organism is not an aggregate of independent traits, each determined by one gene. A "trait" is really an abstraction, a term of convenience in description. One gene may affect many traits (a condition termed pleiotropic). The white gene in *Drosophila* flies is pleiotropic; it affects the colour of the eyes and of the testicular envelope in the males, the fecundity and the shape of the spermatheca in the females, and the longevity of both sexes. In humans many diseases caused by a single defective gene will have a variety of symptoms, all pleiotropic manifestations of the gene.

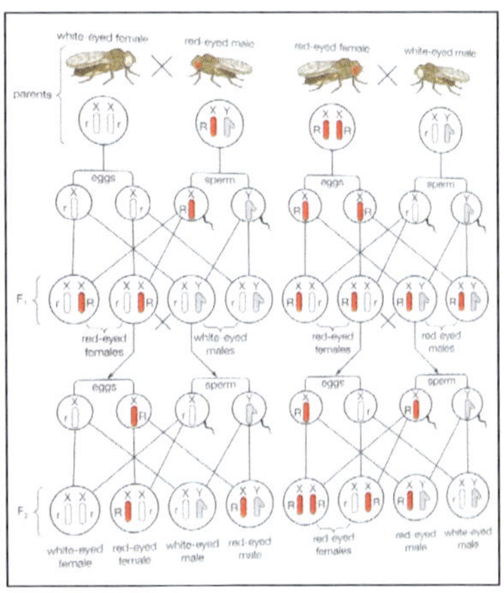

Sex-linked inheritance of white eyes in Drosophila flies.

Allelic Interactions

Dominance Relationships

The operation of Mendelian inheritance is frequently more complex than in the case of the traits recorded by Mendel. In the first place, clear-cut dominance and recessiveness are by no means always found. When red- and white-flowered varieties of four-o'clock plants or snapdragons are crossed, for example, the F_1 hybrids have flowers of intermediate pink or rose colour, a situation that seems more explicable by the blending notion of inheritance than by Mendelian concepts. That the inheritance of flower colour is indeed due to Mendelian mechanisms becomes apparent when the F_1 hybrids are allowed to cross, yielding an F_2 generation of red-, pink-, and white-flowered plants in a ratio of 1 red : 2 pink : 1 white. Obviously the hereditary information for the production of red and white flowers had not been blended away in the first hybrid generation, as flowers of these colours were produced in the second generation of hybrids.

The apparent blending in the F_1 generation is explained by the fact that the gene alleles that govern flower colour in four-o'clocks show an incomplete dominance relationship. Suppose then that a gene allele R_1 is responsible for red flowers and R_2 for white; the homozygotes R_1R_1 and R_2R_2 are red and white respectively, and the heterozygotes R_1R_2 have pink flowers. A similar pattern of lack of dominance is found in Shorthorn cattle. In diverse organisms, dominance ranges from complete (a heterozygote indistinguishable from one of the homozygotes) to incomplete (heterozygotes exactly intermediate) to excessive or overdominance (a heterozygote more extreme than either homozygote).

Another form of dominance is one in which the heterozygote displays the phenotypic characteristics of both alleles. This is called codominance; an example is seen in the MN blood group system of human beings. MN blood type is governed by two alleles, *M* and *N*. Individuals who are homozygous for the Mallele have a surface molecule (called the M antigen) on their red blood cells. Similarly, those homozygous for the *N* allele have the N antigen on the red blood cells. Heterozygotes—those with both alleles—carry both antigens.

Multiple Alleles

The traits discussed so far all have been governed by the interaction of two possible alleles. Many genes, however, are represented by multiple allelic forms within a population. (One individual, of course, can possess only two of these multiple alleles.) Human blood groups—in this case, the well-known ABO system—again provide an example. The gene that governs ABO blood types has three alleles: I^A, I^B, and I^O. I^A and I^B are codominant, but I^O is recessive. Because of the multiple alleles and their various dominance relationships, there are four phenotypic ABO blood types: type A (genotypes I^AI^A and I^AI^O), type B (genotypes I^BI^B and I^BI^O), type AB (genotype I^AI^B), and type O (genotype I^OI^O).

Gene Interactions

Many individual traits are affected by more than one gene. For example, the coat colour in many mammals is determined by numerous genes interacting to produce the result. The great variety of colour patterns in cats, dogs, and other domesticated animals is the result of different combinations of complexly interacting genes. The gradual unraveling of their modes of inheritance was one of the active fields of research in the early years of genetics.

Two or more genes may produce similar and cumulative effects on the same trait. In humans the skin-colour difference between so-called blacks and so-called whites is due to several (probably four or more) interacting pairs of genes, each of which increases or decreases the skin pigmentation by a relatively small amount.

Epistatic Genes

Some genes mask the expression of other genes just as a fully dominant allele masks the expression of its recessive counterpart. A gene that masks the phenotypic effect of another gene is called an epistatic gene; the gene it subordinates is the hypostatic gene. The gene for albinism (lack of pigment) in humans is an epistatic gene. It is not part of the interacting skin-colour genes described above; rather, its dominant allele is necessary for the development of any skin pigment, and its recessive homozygous state results in the albino condition regardless of how many other pigment genes may be present. Albinism thus occurs in some individuals among dark- or intermediate-pigmented peoples as well as among light-pigmented peoples.

The presence of epistatic genes explains much of the variability seen in the expression of such dominantly inherited human diseases as Marfan syndrome and neurofibromatosis. Because of the effects of an epistatic gene, some individuals who inherit a dominant, disease-causing gene show only partial symptoms of the disease; some in fact may show no expression of the disease-causing gene, a condition referred to as nonpenetrance. The individual in whom such a nonpenetrant mutant gene exists will be phenotypically normal but still capable of passing the deleterious gene on to offspring, who may exhibit the full-blown disease.

Examples of epistasis abound in nonhuman organisms. In mice, as in humans, the gene for albinism has two variants: the allele for nonalbino and the allele for albino. The latter allele is unable to synthesize the pigment melanin. Mice, however, have another pair of alleles involved in melanin placement. These are the agouti allele, which produces dark melanization of the hair except for a yellow band at the tip, and the black allele, which produces melanization of the whole hair. If melanin cannot be formed (the situation in the mouse homozygous for the albino gene), neither agouti nor black can be expressed. Hence, homozygosity for the albinism gene is epistatic to the agouti/black alleles and prevents their expression.

Complementation

The phenomenon of complementation is another form of interaction between nonallelic genes. For example, there are mutant genes that in the homozygous state produce profound deafness in humans. One would expect that the children of two persons with such hereditary deafness would be deaf. This is frequently not the case, because the parents' deafness is often caused by different genes. Since the mutant genes are not alleles, the child becomes heterozygous for the two genes and hears normally. In other words, the two mutant genes complement each other in the child. Complementation thus becomes a test for allelism. In the case of congenital deafness cited above, if all the children had been deaf, one could assume that the deafness in each of the parents was owing to mutant genes that were alleles. This would be more likely to occur if the parents were genetically related (consanguineous).

Polygenic Inheritance

The greatest difficulties of analysis and interpretation are presented by the inheritance of many quantitative or continuously varying traits. Inheritance of this kind produces variations in degree rather than in kind, in contrast to the inheritance of discontinuous traits resulting from single genes of major effect. The yield of milk in different breeds of cattle; the egg-laying capacity in poultry; and the stature, shape of the head, blood pressure, and intelligence in humans range in continuous series from one extreme to the other and are significantly dependent on environmental conditions. Crosses of two varieties differing in such characters usually give F_1 hybrids intermediate between the parents. At first sight this situation suggests a blending inheritance through "blood" rather than Mendelian inheritance; in fact, it was probably observations of this kind of inheritance that suggested the folk idea of "blood theory."

It has, however, been shown that these characters are polygenic—i.e., determined by several or many genes, each taken separately producing only a slight effect on the phenotype, as small as or smaller than that caused by environmental influences on the same characters. That Mendelian segregation does take place with polygenes, as with the genes having major effects (sometimes called oligogenes), is shown by the variation among F_2 and further-generation hybrids being usually much greater than that in the F_1 generation. By selecting among the segregating progenies the desired variants—for example, individuals with the greatest yield, the best size, or a desirable behaviour—it is possible to produce new breeds or varieties sometimes exceeding the parental forms. Hybridization and selection are consequently potent methods that have been used for improvement of agricultural plants and animals.

Polygenic inheritance also applies to many of the birth defects (congenital malformations) seen in humans. Although expression of the defect itself may be discontinuous (as in clubfoot, for example), susceptibility to the trait is continuously variable and follows the rules of polygenic inheritance. When a developmental threshold produced by a polygenically inherited susceptibility and a variety of environmental factors is exceeded, the birth defect results.

Heredity and Environment

Preformism and Epigenesis

A notion that was widespread among pioneer biologists in the 18th century was that the fetus, and hence the adult organism that develops from it, is preformed in the sex cells. Some early microscopists even imagined that they saw a tiny homunculus, a diminutive human figure, encased in the human spermatozoon. The development of the individual from the sex cells appeared deceptively simple: it was merely an increase in the size and growth of what was already present in the sex cells. The antithesis of the early preformation theories was theories of epigenesis, which claimed that the sex cells were structureless jelly and contained nothing at all in the way of rudiments of future organisms. The naive early versions of preformation and epigenesis had to be given up when embryologists showed that the embryo develops by a series of complex but orderly and gradual transformations. Darwin's "Provisional Hypothesis of Pangenesis" was distinctly preformistic; Weismann's theory of determinants in the germ plasm, as well as the early ideas about the relations between genes and traits, also tended toward preformism.

Heredity has been defined as a process that results in the progeny's resembling his parents. A further qualification of this definition states that what is inherited is a potential that expresses itself only after interacting with and being modified by environmental factors. In short, all phenotypic expressions have both hereditary and environmental components, the amount of each varying for different traits. Thus, a trait that is primarily hereditary (e.g., skin colour in humans) may be modified by environmental influences (e.g., suntanning). And conversely, a trait sensitive to environmental modifications (e.g., weight in humans) is also genetically conditioned. Organic development is preformistic insofar as a fertilized egg cell contains a genotype that conditions the events that may occur and is epigenetic insofar as a given genotype allows a variety of possible outcomes. These considerations should dispel the reluctance felt by many people to accept the fact that mental as well as physiological and physical traits in humans are genetically conditioned. Genetic conditioning does not mean that heredity is the "dice of destiny." At least in principle, but not invariably in practice, the development of a trait may be manipulated by changes in the environment.

Heritability

Although hereditary diseases and malformations are, unfortunately, by no means uncommon in the aggregate, no one of them occurs very frequently. The characteristics by which one person is distinguished from another—such as facial features, stature, shape of the head, skin, eye and hair colours, and voice—are not usually inherited in a clear-cut Mendelian manner, as are some hereditary malformations and diseases. This is not as strange as it may seem. The kinds of gene changes, or mutations, that produce morphological or physiological effects drastic enough to be clearly set apart from the more usual phenotypes are likely to cause diseases or malformations just because they are so drastic.

The variations that occur among healthy persons are, as a general rule, caused by polygenes with individually small effects. The same is true of individual differences among members of various animal and plant species. Even brown-blue eye colour in humans, which in many families behaves as if caused by two forms of a single gene (brown being dominant and blue recessive), is often blurred by minor gene modifiers of the pigmentation. Some apparently blue-eyed persons actually carry the gene for the brown eye colour, but several additional modifier genes decrease the amount of brown pigment in the iris. This type of genetic process can influence susceptibility to many diseases (e.g., diabetes) or birth defects (e.g., cleft lip—with or without cleft palate).

The question geneticists must often attempt to answer is how much of the observed diversity between persons or between individuals of any species is because of hereditary, or genotypic, variations and how much of it is because of environmental influences. Applied to human beings, this is sometimes referred to as the nature-nurture problem. With animals or plants the problem is evidently more easily soluble than it is with people. Two complementary approaches are possible. First, individual organisms or their progenies are raised in environments as uniform as can be provided, with food, temperature, light, humidity, etc., carefully controlled. The differences that persist between such individuals or progenies probably reflect genotypic differences. Second, individuals with similar or identical genotypes are placed in different environments. The phenotypic differences may then be ascribed to environmental induction. Experiments combining both approaches have been carried out on several species of plants that grow naturally at different altitudes, from sea level to the alpine zone of the Sierra Nevada in California. Young yarrow plants (Achillea) were cut into three parts, and the cuttings were replanted in experimental gardens at sea level, at mid-altitude (4,800 feet [1,460 metres]), and at high altitude (10,000 feet [3,050 metres]). It was observed that the plants native at sea level grow best in their native habitat, grow less well at mid-altitudes, and die at high altitudes. On the other hand, the alpine race survives and develops better at the high-altitude transplant station than it does at lower altitudes.

With organisms that cannot survive being cut into pieces and placed in controlled environments, a partitioning of the observed variability into genetic and environmental components may be attempted by other methods. Suppose that in a certain population individuals vary in stature, weight, or some other trait. These characters can be measured in many pairs of parents and in their progenies raised under different environmental conditions. If the variation is owing entirely to environment and not at all to heredity, then the expression of the character in the parents and in the offspring will show no correlation (heritability = zero). On the other hand, if the environment is unimportant and the character is uncomplicated by dominance, then the means of this character in the progenies will be the same as the means of the parents; with differences in the expression in females and in males taken into account, the heritability will equal unity. In reality, most heritabilities are found to lie between zero and one.

It is important to understand clearly the meaning of heritability estimates. They show that, given the range of the environments in which the experimental animals lived, one could predict the average body sizes in the progenies of pigs better than one could predict the average numbers of piglets in a litter. The heritability is, however, not an inherent or unchangeable property of each character. If one could make the environments more uniform, the heritabilities would rise, and with more-diversified environments they would decrease. Similarly, in populations that are more variable genetically, the heritabilities increase, and in genetically uniform ones, they decrease. In humans the situation is even more complex, because the environments of the parents and of their children are in many ways interdependent. Suppose, for example, that one wishes to study the heritability of stature, weight, or susceptibility to tuberculosis. The stature, weight, and liability to contract tuberculosis depend to some extent on the quality of nutrition and generally on the economic well-being of the family. If no allowance is made for this fact, the heritability estimates arrived at may be spurious; such heritabilities have indeed been claimed for such things as administrative, legal, or military talents and for social eminence in general. It is evident that having socially eminent parents makes it easier for the children to achieve such eminence also; biological heredity may have little or nothing to do with this.

A general conclusion from the evidence now available may be stated as follows: diversity in almost any trait—physical, physiological, or behavioural—owes in part to genetic variables and in part to environmental variables. In any array of environments, individuals with more nearly similar genetic endowments are likely to show a greater average resemblance than the carriers of more diverse genetic endowments. It is, however, also true that in different environments the carriers of similar genetic endowments may grow, develop, and behave in different ways.

Chromosome

Chromosomes are thread-like molecules that carry hereditary information for everything from height to eye color. They are made of protein and one molecule of DNA, which contains an organism's genetic instructions, passed down from parents. In humans, animals, and plants, most chromosomes are arranged in pairs within the nucleus of a cell. Humans have 22 of these chromosome pairs, called autosomes.

Sex Determination

Humans have an additional pair of sex chromosomes for a total of 46 chromosomes. The sex chromosomes are referred to as X and Y, and their combination determines a person's sex. Typically, human females have two X chromosomes while males possess an XY pairing. This XY sex-determination system is found in most mammals as well as some reptiles and plants.

Whether a person has XX or XY chromosomes is determined when a sperm fertilizes an egg. Unlike the body's other cells, the cells in the egg and sperm – called gametes or sex cells – possess only one chromosome. Gametes are produced by meiosis cell division, which results in the divided cells having half the number of chromosomes as the parent, or progenitor, cells. In the case of humans, this means that parent cells have two chromosomes and gametes have one.

All of the gametes in the mother's eggs possess X chromosomes. The father's sperm contains about half X and half Y chromosomes. The sperm are the variable factor in determining the sex of the baby. If the sperm carries an X chromosome, it will combine with the egg's X chromosome to form a female zygote. If the sperm carries a Y chromosome, it will result in a male.

During fertilization, gametes from the sperm combine with gametes from the egg to form a zygote. The zygote contains two sets of 23 chromosomes, for the required 46. Most women are 46XX and most men are 46XY.

There are some variations, though. A few births out of a thousand of babies are born with a single sex chromosome (45X or 45Y) and are referred to as sex monosomies. Others are born with three or more sex chromosomes (47XXX, 47XYY or 47XXY, etc.) and are called sex polysomies. "In addition, some males are born 46XX due to the translocation of a tiny section of the sex determining region of the Y chromosome," said WHO. "Similarly some females are also born 46XY due to mutations in the Y chromosome. Clearly, there are not only females who are XX and males who are XY, but rather, there is a range of chromosome complements, hormone balances, and phenotypic variations that determine sex."

It is important to remember that sex and gender have two separate definitions and many cultures include more labels than simply "male" and "female" to identify others.

The Structure of X and Y Chromosomes

While the chromosomes for other parts of the body are the same size and shape - forming an identical pairing - the X and Y chromosomes have different structures.

The X chromosome is significantly longer than the Y chromosome and contains hundreds more genes. Because the additional genes in the X chromosome have no counterpart in the Y chromosome, the X genes are dominant. This means that almost any gene on the X, even if it is recessive in the female, will be expressed in males. These are referred to as X-linked genes. Genes found only on the Y chromosome are referred to as Y-linked genes, and expressed only in males. Genes on either sex chromosome can be called sex-linked genes.

There are approximately 1,098 X-linked genes, though most of them are not for female anatomical characteristics. In fact, many are linked to disorders such as hemophilia, Duchenne muscular dystrophy, fragile-X syndrome and several others. They are responsible for red-green color blindness, considered the most common genetic disorder and found most often in males. The non-sex feature X-linked genes are also responsible for male pattern baldness.

In contrast to the large X chromosome, the Y chromosome contains only 26 genes. Sixteen of these genes are responsible for cell maintenance. Nine are involved in sperm production, and if some are missing or defective, low sperm counts or infertility may occur. One gene, called the SRY gene, is responsible for male sexual traits. The SRY gene triggers the activation and regulation of another gene, found on a non-sex chromosome, called the Sox9. The Sox9 triggers the development of non-sexed gonads into testes instead of ovaries.

Sex Chromosome Abnormalities

Abnormalities in the sex chromosome combination can result in a variety of gender-specific conditions that are rarely lethal.

Female abnormalities result in Turner syndrome or Trisomy X. Turner syndrome occurs when females have only one X chromosome instead of two. Symptoms include failure of the sex organs to normally mature, which may lead to infertility, small breasts and no menstruation; short stature; a wide, shield-shaped chest; and a wide, webbed neck.

Trisomy X syndrome is caused by three X chromosomes instead of two. Symptoms include tall stature, speech delays, premature ovarian failure or ovarian abnormalities, and weak muscle tone-although many girls and women exhibit no symptoms.

Males can be affected by Klinefelter syndrome. Symptoms include breast development, abnormal proportions such as large hips, tall height, infertility, and small testicles.

Chromosomes and Genes

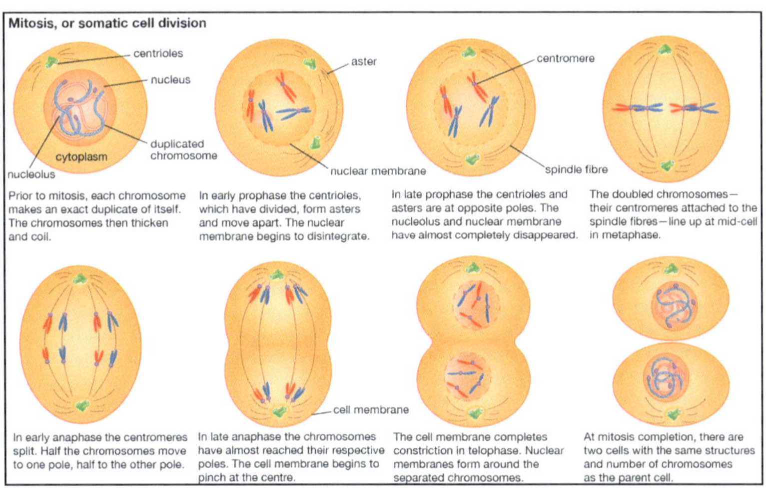

Each individual in a sexually reproducing species inherits two alleles for each gene, one from each parent. Furthermore, when such an individual forms sex cells, each of the resultant gametes receives one member of each allelic pair. The formation of gametes occurs through a process of cell division called meiosis. When gametes unite in fertilization, the double dose of hereditary material is restored, and a new individual is created. This individual, consisting at first of only one cell, grows via mitosis, a process of repeated cell divisions. Mitosis differs from meiosis in that each daughter cell receives a full copy of all the hereditary material found in the parent cell.

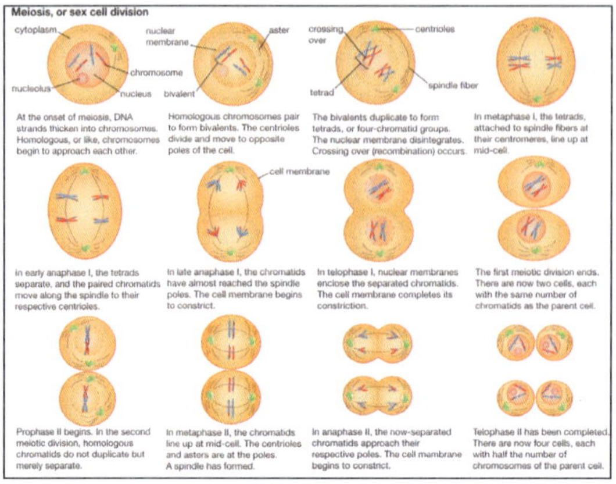

The formation of gametes (sex cells) occurs during the process of meiosis.

It is apparent that the genes must physically reside in cellular structures that meet two criteria. First, these structures must be replicated and passed on to each generation of daughter cells during mitosis. Second, they must be organized into homologous pairs, one member of which is parceled out to each gamete formed during meiosis.

As early as 1848, biologists had observed that cell nuclei resolve themselves into small rodlike bodies during mitosis; later these structures were found to absorb certain dyes and so came to be called chromosomes (coloured bodies). During the early years of the 20th century, cellular studies using ordinary light microscopes clarified the behaviour of chromosomes during mitosis and meiosis, which led to the conclusion that chromosomes are the carriers of genes.

Behavior of Chromosomes during Cell Division

Mitosis

Mitosis is a type of cell division in which one cell (the mother) divides to produce two new cells (the daughters) that are genetically identical to itself. In the context of the cell cycle, mitosis is the part of the division process in which the DNA of the cell's nucleus is split into two equal sets of chromosomes.

The great majority of the cell divisions that happen in your body involve mitosis. During development and growth, mitosis populates an organism's body with cells, and throughout an organism's life, it replaces old, worn-out cells with new ones. For single-celled eukaryotes like yeast, mitotic divisions are actually a form of reproduction, adding new individuals to the population.

In all of these cases, the "goal" of mitosis is to make sure that each daughter cell gets a perfect, full set of chromosomes. Cells with too few or too many chromosomes usually don't function well: they may not survive, or they may even cause cancer. So, when cells undergo mitosis, they don't just divide their DNA at random and toss it into piles for the two daughter cells. Instead, they split up their duplicated chromosomes in a carefully organized series of steps.

Phases of Mitosis

Mitosis consists of four basic phases: prophase, metaphase, anaphase, and telophase. Some textbooks list five, breaking prophase into an early phase (called prophase) and a late phase (called prometaphase). These phases occur in strict sequential order, and cytokinesis - the process of dividing the cell contents to make two new cells - starts in anaphase or telophase.

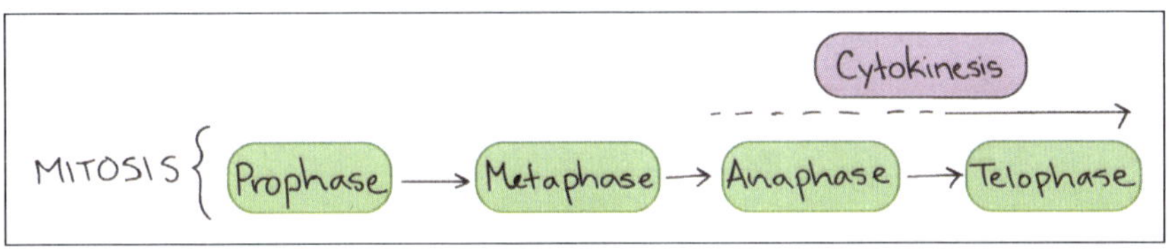

We can remember the order of the phases with the famous mnemonic: [Please] Pee on the MAT. But don't get too hung up on names – what's most important to understand is what's happening at each stage, and why it's important for the division of the chromosomes.

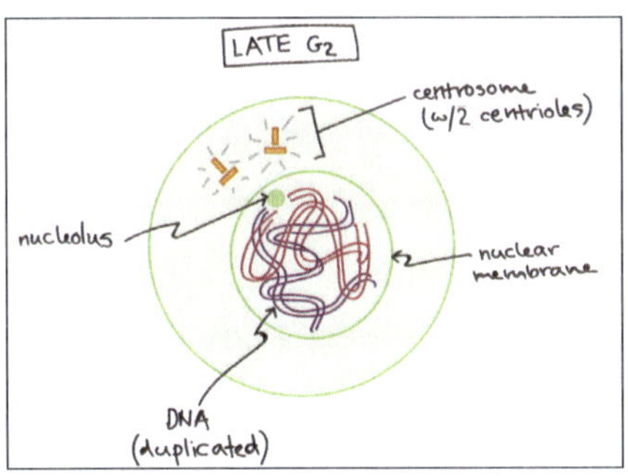

Let's start by looking at a cell right before it begins mitosis. This cell is in interphase (late G_2 phase) and has already copied its DNA, so the chromosomes in the nucleus each consist of two connected copies, called sister chromatids. You can't see the chromosomes very clearly at this point, because they are still in their long, stringy, decondensed form.

This animal cell has also made a copy of its centrosome, an organelle that will play a key role in orchestrating mitosis, so there are two centrosomes. (Plant cells generally don't have centrosomes with centrioles, but have a different type of microtubule organizing center that plays a similar role).

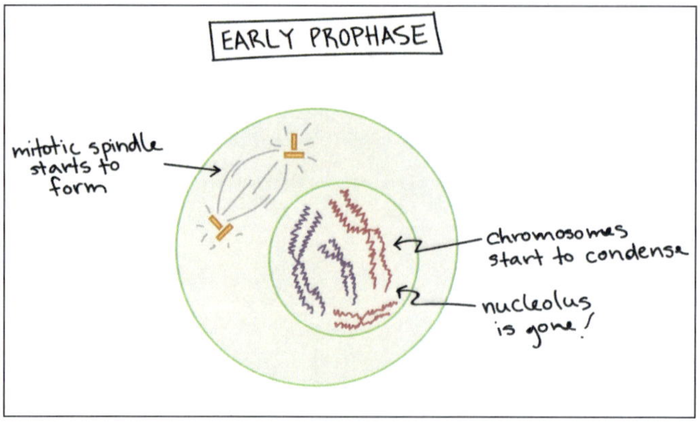

In early prophase, the cell starts to break down some structures and build others up, setting the stage for division of the chromosomes.

- The chromosomes start to condense (making them easier to pull apart later on).
- The mitotic spindle begins to form. The spindle is a structure made of microtubules, strong fibers that are part of the cell's "skeleton." Its job is to organize the chromosomes and move them around during mitosis. The spindle grows between the centrosomes as they move apart.
- The nucleolus (or nucleoli, plural), a part of the nucleus where ribosomes are made, disappears. This is a sign that the nucleus is getting ready to break down.

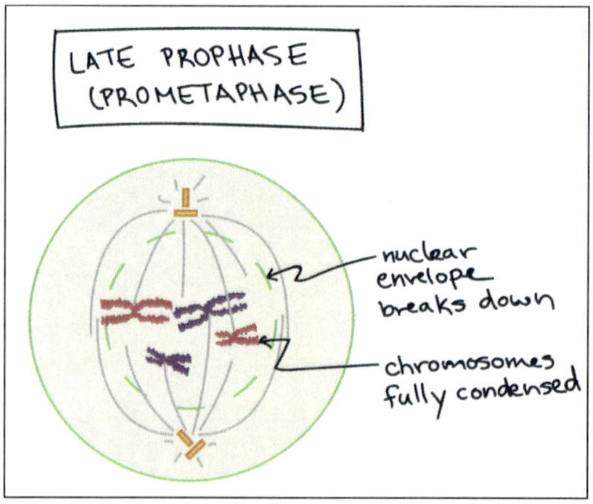

In late prophase (sometimes also called prometaphase), the mitotic spindle begins to capture and organize the chromosomes.

- The chromosomes finish condensing, so they are very compact.
- The nuclear envelope breaks down, releasing the chromosomes.
- The mitotic spindle grows more, and some of the microtubules start to "capture" chromosomes.

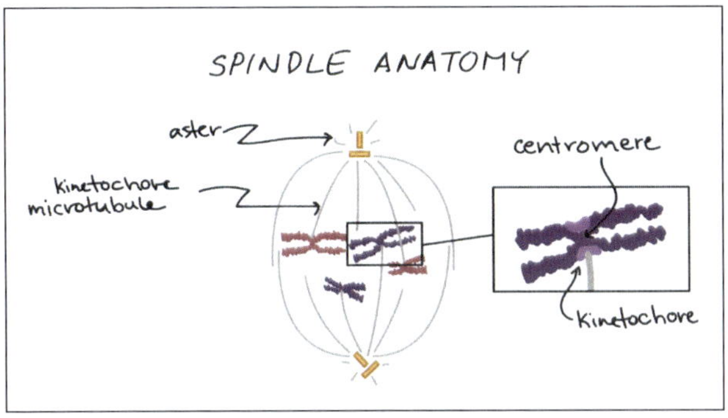

Microtubules can bind to chromosomes at the kinetochore, a patch of protein found on the centromere of each sister chromatid. (Centromeres are the regions of DNA where the sister chromatids are most tightly connected.)

Microtubules that bind a chromosome are called kinetochore microtubules. Microtubules that don't bind to kinetochores can grab on to microtubules from the opposite pole, stabilizing the spindle. More microtubules extend from each centrosome towards the edge of the cell, forming a structure called the aster.

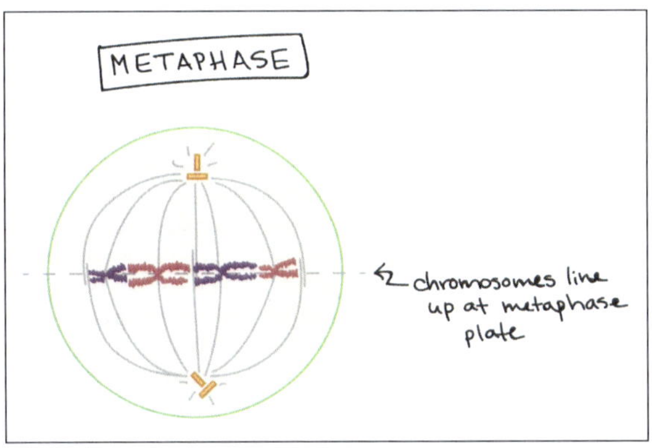

In metaphase, the spindle has captured all the chromosomes and lined them up at the middle of the cell, ready to divide.

- All the chromosomes align at the metaphase plate (not a physical structure, just a term for the plane where the chromosomes line up).
- At this stage, the two kinetochores of each chromosome should be attached to microtubules from opposite spindle poles.

Before proceeding to anaphase, the cell will check to make sure that all the chromosomes are at the metaphase plate with their kinetochores correctly attached to microtubules. This is called the spindle checkpoint and helps ensure that the sister chromatids will split evenly between the two daughter cells when they separate in the next step. If a chromosome is not properly aligned or attached, the cell will halt division until the problem is fixed.

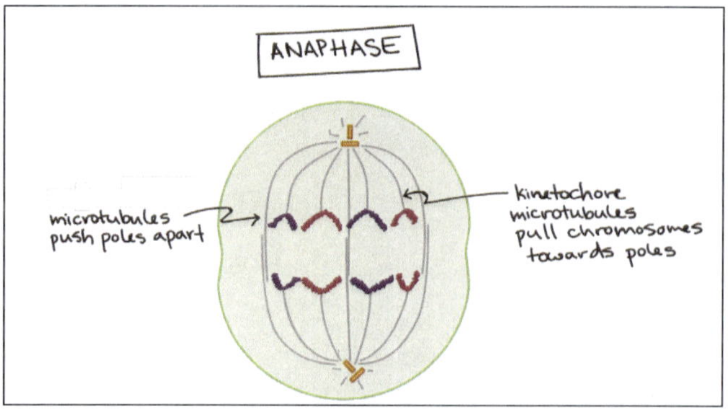

In anaphase, the sister chromatids separate from each other and are pulled towards opposite ends of the cell.

- The protein "glue" that holds the sister chromatids together is broken down, allowing them to separate. Each is now its own chromosome. The chromosomes of each pair are pulled towards opposite ends of the cell.
- Microtubules not attached to chromosomes elongate and push apart, separating the poles and making the cell longer.

All of these processes are driven by motor proteins, molecular machines that can "walk" along microtubule tracks and carry a cargo. In mitosis, motor proteins carry chromosomes or other microtubules as they walk.

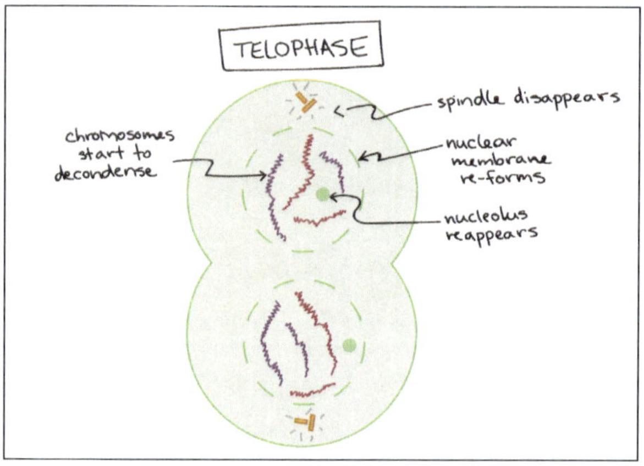

In telophase, the cell is nearly done dividing, and it starts to re-establish its normal structures as cytokinesis (division of the cell contents) takes place.

- The mitotic spindle is broken down into its building blocks.
- Two new nuclei form, one for each set of chromosomes. Nuclear membranes and nucleoli reappear.
- The chromosomes begin to decondense and return to their "stringy" form.

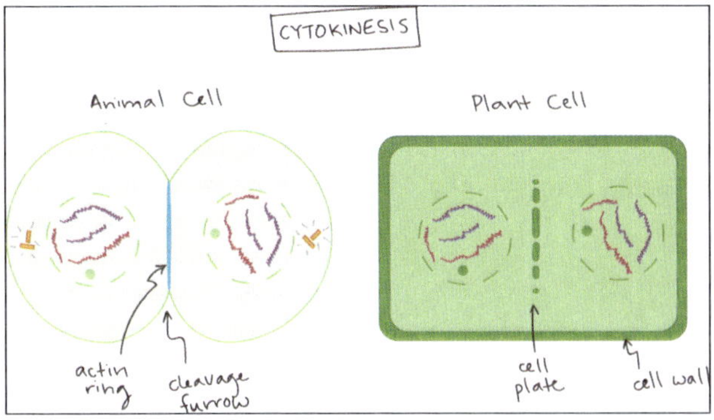

Cytokinesis, the division of the cytoplasm to form two new cells, overlaps with the final stages of mitosis. It may start in either anaphase or telophase, depending on the cell, and finishes shortly after telophase.

In animal cells, cytokinesis is contractile, pinching the cell in two like a coin purse with a drawstring. The "drawstring" is a band of filaments made of a protein called actin, and the pinch crease is known as the cleavage furrow. Plant cells can't be divided like this because they have a cell wall and are too stiff. Instead, a structure called the cell plate forms down the middle of the cell, splitting it into two daughter cells separated by a new wall.

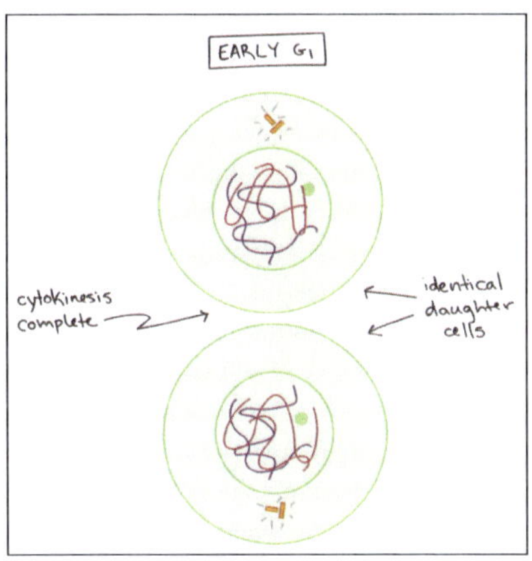

When cytokinesis finishes, we end up with two new cells, each with a complete set of chromosomes identical to those of the mother cell. The daughter cells can now begin their own cellular "lives," and – depending on what they decide to be when they grow up – may undergo mitosis themselves, repeating the cycle.

Meiosis

Meiosis is used for just one purpose in the human body: the production of gametes-sex cells, or sperm and eggs. Its goal is to make daughter cells with exactly half as many chromosomes as the starting cell.

To put that another way, meiosis in humans is a division process that takes us from a diploid cell-one with two sets of chromosomes-to haploid cells-ones with a single set of chromosomes. In humans, the haploid cells made in meiosis are sperm and eggs. When a sperm and an egg join in fertilization, the two haploid sets of chromosomes form a complete diploid set: a new genome.

Phases of Meiosis

In many ways, meiosis is a lot like mitosis. The cell goes through similar stages and uses similar strategies to organize and separate chromosomes. In meiosis, however, the cell has a more complex task. It still needs to separate sister chromatids (the two halves of a duplicated chromosome), as in mitosis. But it must also separate homologous chromosomes, the similar but nonidentical chromosome pairs an organism receives from its two parents.

These goals are accomplished in meiosis using a two-step division process. Homologue pairs separate during a first round of cell division, called meiosis I. Sister chromatids separate during a second round, called meiosis II.

Since cell division occurs twice during meiosis, one starting cell can produce four gametes (eggs or sperm). In each round of division, cells go through four stages: prophase, metaphase, anaphase, and telophase.

Meiosis I

Before entering meiosis I, a cell must first go through interphase. As in mitosis, the cell grows during G_1 phase, copies all of its chromosomes during S phase, and prepares for division during G_2 phase.

During prophase I, differences from mitosis begin to appear. As in mitosis, the chromosomes begin to condense, but in meiosis I, they also pair up. Each chromosome carefully aligns with its homologue partner so that the two match up at corresponding positions along their full length.

For instance, in the image below, the letters A, B, and C represent genes found at particular spots on the chromosome, with capital and lowercase letters for different forms, or alleles, of each gene. The DNA is broken at the same spot on each homologue—here, between genes B and C—and reconnected in a criss-cross pattern so that the homologues exchange part of their DNA.

This process, in which homologous chromosomes trade parts, is called crossing over. It's helped along by a protein structure called the synaptonemal complex that holds the homologues together. The chromosomes would actually be positioned one on top of the other—as in the image below—throughout crossing over; they're only shown side-by-side in the image above so that it's easier to see the exchange of genetic material.

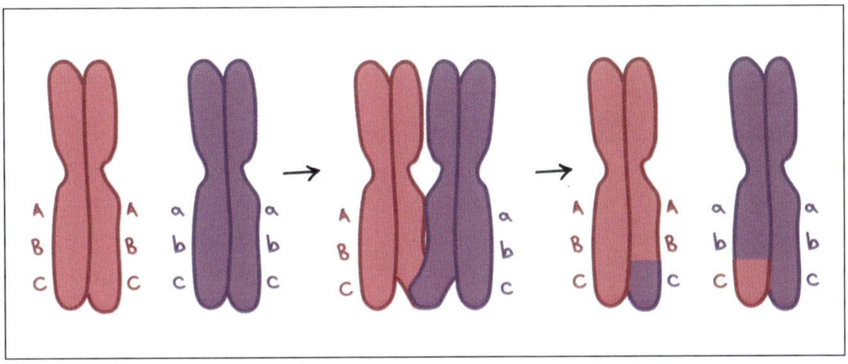

The process of meiosis.

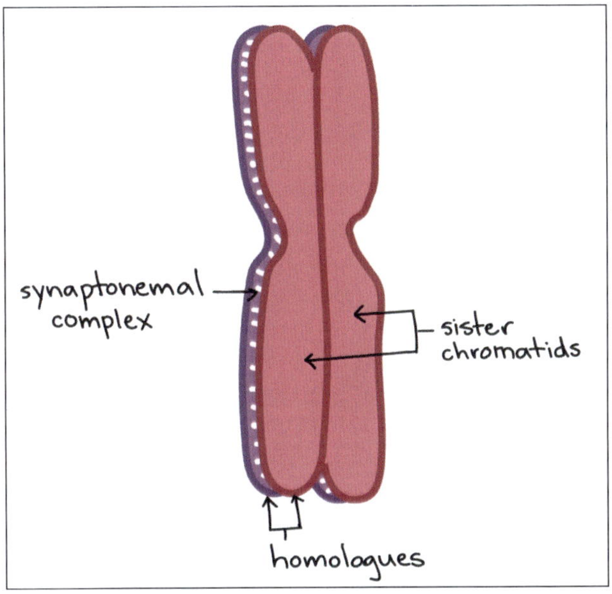

The process of meiosis.

You can see crossovers under a microscope as chiasmata, cross-shaped structures where homologues are linked together. Chiasmata keep the homologues connected to each other after the synaptonemal complex breaks down, so each homologous pair needs at least one. It's common for multiple crossovers (up to 25!) to take place for each homologue pair.

The spots where crossovers happen are more or less random, leading to the formation of new, "remixed" chromosomes with unique combinations of alleles.

After crossing over, the spindle begins to capture chromosomes and move them towards the center of the cell (metaphase plate). This may seem familiar from mitosis, but there is a twist. Each chromosome attaches to microtubules from just one pole of the spindle, and the two homologues of a pair bind to microtubules from opposite poles. So, during metaphase I, homologue pairs—not individual chromosomes—line up at the metaphase plate for separation.

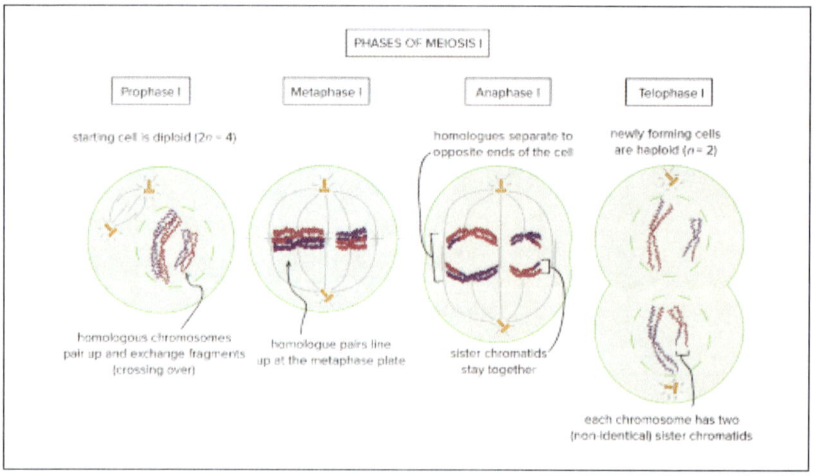

When the homologous pairs line up at the metaphase plate, the orientation of each pair is random. For instance, in the diagram above, the pink version of the big chromosome and the purple version of the little chromosome happen to be positioned towards the same pole and go into the same cell. But the orientation could have equally well been flipped, so that both purple chromosomes went into the cell together. This allows for the formation of gametes with different sets of homologues.

In anaphase I, the homologues are pulled apart and move apart to opposite ends of the cell. The sister chromatids of each chromosome, however, remain attached to one another and don't come apart.

Finally, in telophase I, the chromosomes arrive at opposite poles of the cell. In some organisms, the nuclear membrane re-forms and the chromosomes decondense, although in others, this step is skipped—since cells will soon go through another round of division, meiosis II. Cytokinesis usually occurs at the same time as telophase I, forming two haploid daughter cells.

Meiosis II

Cells move from meiosis I to meiosis II without copying their DNA. Meiosis II is a shorter and simpler process than meiosis I, and you may find it helpful to think of meiosis II as "mitosis for haploid cells."

The cells that enter meiosis II are the ones made in meiosis I. These cells are haploid—have just one chromosome from each homologue pair—but their chromosomes still consist of two sister chromatids. In meiosis II, the sister chromatids separate, making haploid cells with non-duplicated chromosomes.

During prophase II, chromosomes condense and the nuclear envelope breaks down, if needed. The centrosomes move apart, the spindle forms between them, and the spindle microtubules begin to capture chromosomes.

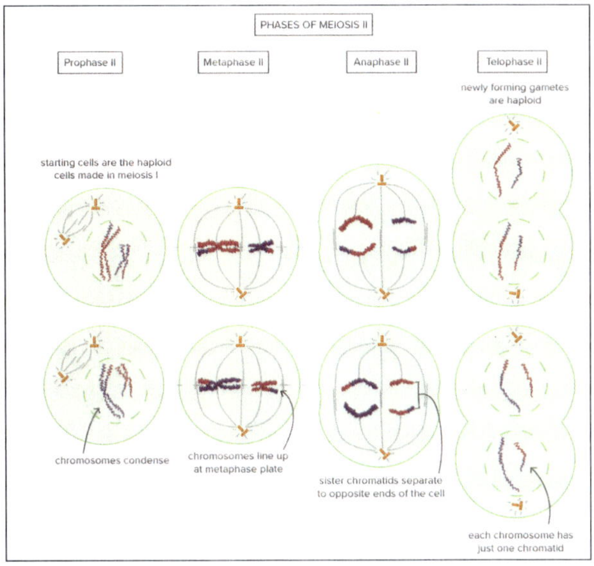

The two sister chromatids of each chromosome are captured by microtubules from opposite spindle poles. In metaphase II, the chromosomes line up individually along the metaphase plate. In anaphase II, the sister chromatids separate and are pulled towards opposite poles of the cell.

In telophase II, nuclear membranes form around each set of chromosomes, and the chromosomes decondense. Cytokinesis splits the chromosome sets into new cells, forming the final products of meiosis: four haploid cells in which each chromosome has just one chromatid. In humans, the products of meiosis are sperm or egg cells.

Mixing and Matching of Genes

The gametes produced in meiosis are all haploid, but they're not genetically identical. For example, take a look the meiosis II diagram, which shows the products of meiosis for a cell with 2n = 4, n, equals, 4 chromosomes. Each gamete has a unique "sample" of the genetic material present in the starting cell.

As it turns out, there are many more potential gamete types than just the four shown in the diagram, even for a cell with only four chromosomes. The two main reasons we can get many genetically different gametes are:

- Crossing over: The points where homologues cross over and exchange genetic material are chosen more or less at random, and they will be different in each cell that goes through meiosis. If meiosis happens many times, as in humans, crossovers will happen at many different points.
- Random orientation of homologue pairs: The random orientation of homologue pairs in metaphase I allows for the production of gametes with many different assortments of homologous chromosomes.

In a human cell, the random orientation of homologue pairs alone allows for over 8 million different types of possible gametes. When we layer crossing over on top of this, the number of genetically different gametes that you—or any other person—can make is effectively infinite.

Chromosome Mutations

A chromosome mutation is an unpredictable change that occurs in a chromosome. These changes are most often brought on by problems that occur during meiosis (division process of gametes) or by mutagens (chemicals, radiation, etc.). Chromosome mutations can result in changes in the number of chromosomes in a cell or changes in the structure of a chromosome. Unlike a gene mutation which alters a single gene or larger segment of DNA on a chromosome, chromosome mutations change and impact the entire chromosome.

Chromosome Structure

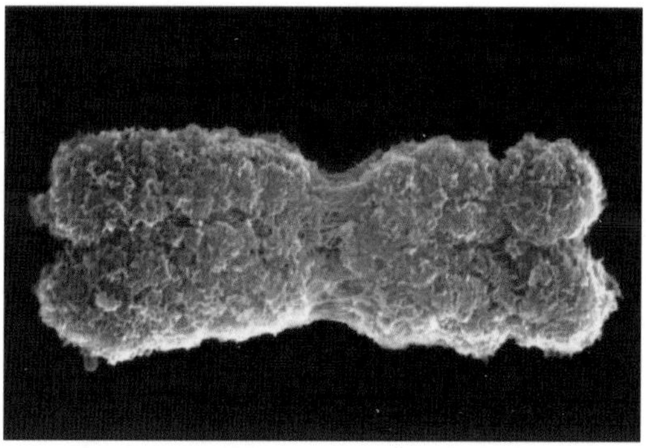

Eukaryotic Nuclear Chromosome.

Chromosomes are long, stringy aggregates of genes that carry heredity information (DNA). They are formed from chromatin, a mass of genetic material consisting of DNA that is tightly coiled around proteins called histones. Chromosomes are located in the nucleus of our cells and condense prior to the process of cell division. A non-duplicated chromosome is single-stranded and is comprised of a centromere region that connects two arm regions. The short arm region is called the p arm and the long arm region is called the q arm.

In preparation for the division of the nucleus, chromosomes must be duplicated to ensure that the resulting daughter cells end up with the appropriate number of chromosomes. An identical copy of each chromosome is therefore produced through DNA replication. Each duplicated chromosome is comprised of two identical chromosomes called sister chromatids that are connected at the centromere region. Sister chromatids separate prior to the completion of cell division.

Chromosome Structure Changes

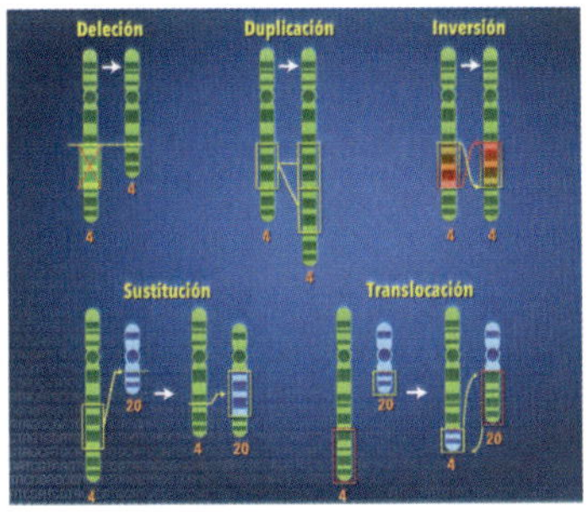

Different types of chromosomal abnormalities.

Duplications and breakages of chromosomes are responsible for a type of chromosome mutation that alters chromosome structure. These changes affect protein production by changing the genes on the chromosome. Chromosome structure changes are often harmful to an individual leading to developmental difficulties and even death. Some changes are not as harmful and may have no significant effect on an individual. There are several types of chromosome structure changes that can occur. Some of them include:

- Translocation: The joining of a fragmented chromosome to a non-homologous chromosome is a translocation. The piece of chromosome detaches from one chromosome and moves to a new position on another chromosome.
- Deletion: This mutation results from the breakage of a chromosome in which the genetic material becomes lost during cell division. The genetic material can break off from anywhere on the chromosome.
- Duplication: Duplications are produced when extra copies of genes are generated on a chromosome.
- Inversion: In an inversion, the broken chromosome segment is reversed and inserted back into the chromosome. If the inversion encompasses the centromere of the chromosome, it is called a pericentric inversion. If it involves the long or short arm of the chromosome and does not include the centromere, it is called a paracentric inversion.
- Isochromosome: This type of chromosome is produced by the improper division of the centromere. Isochromosomes contain either two short arms or two long arms. A typical chromosome contains one short arm and one long arm.

Chromosome Number Changes

A chromosome mutation that causes individuals to have an abnormal number of chromosomes is termed aneuploidy. Aneuploid cells occur as a result of chromosome breakage or nondisjunction errors that happen during meiosis or mitosis. Nondisjunction is the failure of homologous chromosomes to separate properly during cell division. It produces individuals with either extra or missing chromosomes. Sex chromosome abnormalities that result from nondisjunction can lead to conditions such as Klinefelter and Turner syndromes. In Klinefelter syndrome, males have one or more extra X sex chromosomes. In Turner syndrome, females have only one X sex chromosome. Down syndrome is an example of a condition that occurs due to nondisjunction in autosomal (non-sex) cells. Individuals with Down syndrome have an extra chromosome on autosomal chromosome 21.

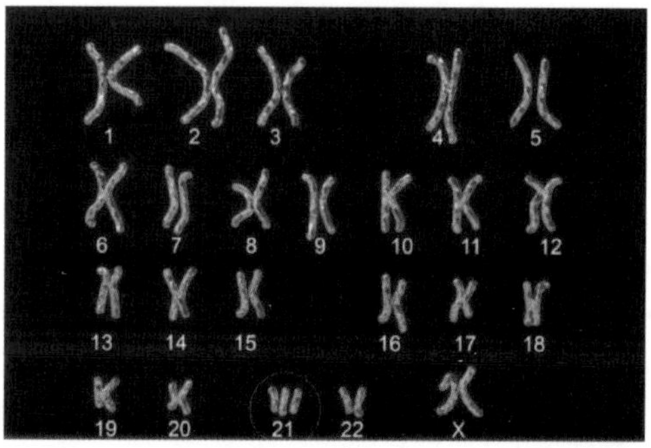

Down syndrome is caused by a chromosomal anomaly: the 21st set having three rather than the normal two chromosomes.

A chromosome mutation that results in individuals with more than one haploid set of chromosomes in a cell is termed polyploidy. A haploid cell is a cell that contains one complete set of chromosomes. Our sex cells are considered haploid and contain 1 complete set of 23 chromosomes. Our autosomal cells are diploid and contain 2 complete sets of 23 chromosomes. If a mutation causes a cell to have three haploid sets, it is called triploidy. If the cell has four haploid sets, it is called tetraploidy.

Sex-Linked Mutations

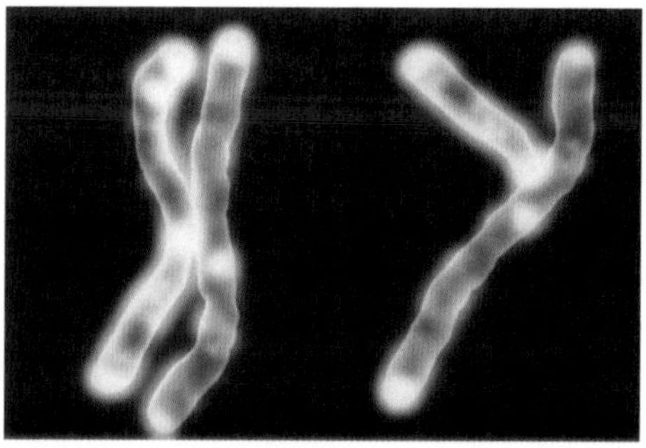

Conceptual representation of the X and Y chromosomes of a human male (Here the Y chromosome (at right) has been altered in shape and size to appear much larger and more Y-shaped than it really is).

Mutations can occur on genes located on sex chromosomes known as sex-linked genes. These genes on either the X chromosome or the Y chromosome determine the genetic characteristics of sex-linked traits. A gene mutation that occurs on the X chromosome can be dominant or recessive. X-linked dominant disorders are expressed in both males and females. X-linked recessive disorders are expressed in males and can be masked in females if the female's second X chromosome is normal. Y chromosome linked disorders are expressed only in males.

Mendelian Inheritance

Mendelian inheritance, also called Mendelism refers to the principles of heredity formulated by Austrian-born botanist, teacher, and Augustinian prelate Gregor Mendel in 1865. These principles compose what is known as the system of particulate inheritance by units, or genes. The later discovery of chromosomes as the carriers of genetic units supported Mendel's two basic laws, known as the law of segregation and the law of independent assortment.

In modern terms, the first of Mendel's laws states that genes are transferred as separate and distinct units from one generation to the next. The two members (alleles) of a gene pair, one on each of paired chromosomes, separate during the formation of sex cells by a parent organism. One-half of the sex cells will have one form of the gene, one-half the other form; the offspring that result from these sex cells will reflect those proportions.

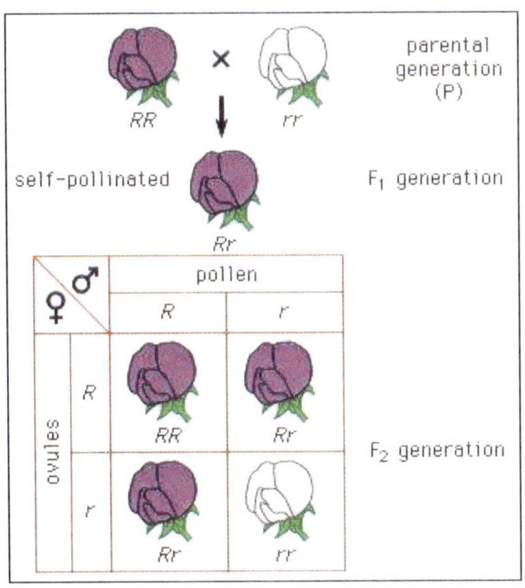

Mendel's law of segregation: Cross of a purple-flowered and a white-flowered strain of peas. R stands for the gene for purple flowers and r for the gene for white flowers.

A modern formulation of the second law, the law of independent assortment, is that the alleles of a gene pair located on one pair of chromosomes are inherited independently of the alleles of a gene pair located on another chromosome pair and that the sex cells containing various assortments of these genes fuse at random with the sex cells produced by the other parent.

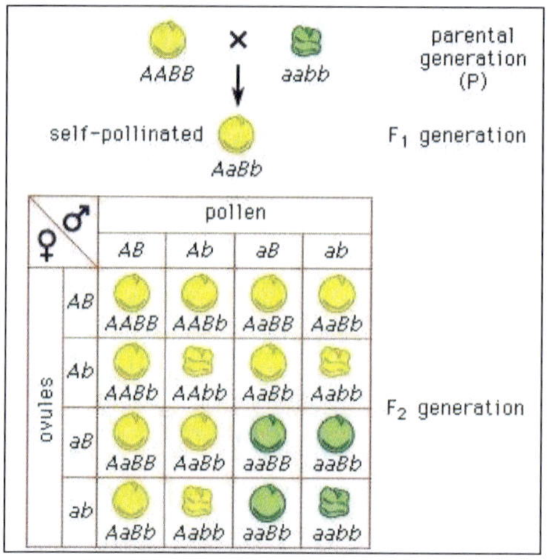

Mendel's law of independent assortment.

In the above figure, the example here shows a cross of peas having yellow and smooth seeds with peas having green and wrinkled seeds. A stands for the gene for yellow and a for the gene for green; B stands for the gene for a smooth surface and b for the gene for a wrinkled surface.

Mendel also developed the law of dominance, in which one allele exerts greater influence than the other on the same inherited character. Mendel developed the concept of dominance from his experiments with plants, based on the supposition that each plant carried two trait units, one of which dominated the other.

References

- Heredity-genetics, science: britannica.com, Retrieved 28 February, 2019
- 27248-chromosomes: livescience.com, Retrieved 1 May, 2019
- The-physical-basis-of-heredity, heredity-genetics, science: britannica.com, Retrieved 10 March, 2019
- Phases-of-mitosis, mitosis, cellular-molecular-biology, biology, science: khanacademy.org, Retrieved 12 August, 2019
- Phases-of-meiosis, meiosis, cellular-molecular-biology, biology, science: khanacademy.org, Retrieved 15 June, 2019
- Chromosome-mutation: thoughtco.com, Retrieved 21 April, 2019
- Mendelism-genetics, science: britannica.com, Retrieved 25 July, 2019

Chapter 3

DNA and RNA: Structure, Function and Processes

DNA is a molecule which is made up of two chains that coil around each other in a double helix formation. RNA is also a chain of nucleotides but exists as a single strand. The topics elaborated in this chapter will help in gaining a better perspective about the different structures, functions and processes associated with DNA and RNA.

DNA

DNA is the information molecule. It stores instructions for making other large molecules, called proteins. These instructions are stored inside each of your cells, distributed among 46 long structures called chromosomes. These chromosomes are made up of thousands of shorter segments of DNA, called genes. Each gene stores the directions for making protein fragments, whole proteins, or multiple specific proteins.

DNA is well-suited to perform this biological function because of its molecular structure, and because of the development of a series of high performance enzymes that are fine-tuned to interact with this molecular structure in specific ways. The match between DNA structure and the activities of these enzymes is so effective and well-refined that DNA has become, over evolutionary time, the universal information-storage molecule for all forms of life. Nature has yet to find a better solution than DNA for storing, expressing, and passing along instructions for making proteins.

Molecular Structure of DNA

In order to understand the biological function of DNA, you first need to understand its molecular structure. This requires learning the vocabulary for talking about the building blocks of DNA, and how these building blocks are assembled to make DNA molecules.

DNA Molecules are Polymers

Polymers are large molecules that are built up by repeatedly linking together smaller molecules, called monomers. Think of how a freight train is built by linking lots of individual boxcars together, or how this sentence is built by sticking together a specific sequence of individual letters (plus spaces and punctuation). In all three cases, the large structure—a train, a sentence, a DNA molecule—is composed of smaller structures that are linked together in non-random sequences— boxcars, letters, and, in the biological case, DNA monomers.

DNA Monomers are called Nucleotides

Just like a sentence "polymer" is composed of letter "monomers," a DNA polymer is composed of monomers called *nucleotides*. A molecule of DNA is a bunch of nucleotide monomers, joined one after another into a very long chain.

Types of Nucleotide Monomers

There are four nucleotide monomers: The English language has a 26 letter alphabet. In contrast, the DNA "alphabet" has only four "letters," the four nucleotide monomers. They have short and easy to remember names: A, C, T, G. Each nucleotide monomer is built from three simple molecular parts: a sugar, a phosphate group, and a nucleobase. (Don't confuse this use of "base" with the other one, which refers to a molecule that raises the pH of a solution; they're two different things).

Sugar and Acid in all Four Monomers are the Same

All four nucleotides (A, T, G and C) are made by sticking a phosphate group and a nucleobase to a sugar. The sugar in all four nucleotides is called deoxyribose. It's a cyclical molecule—most of its atoms are arranged in a ring-structure. The ring contains one oxygen and four carbons. A fifth carbon atom is attached to the fourth carbon of the ring. Deoxyribose also contains a hydroxyl group (-OH) attached to the third carbon in the ring.

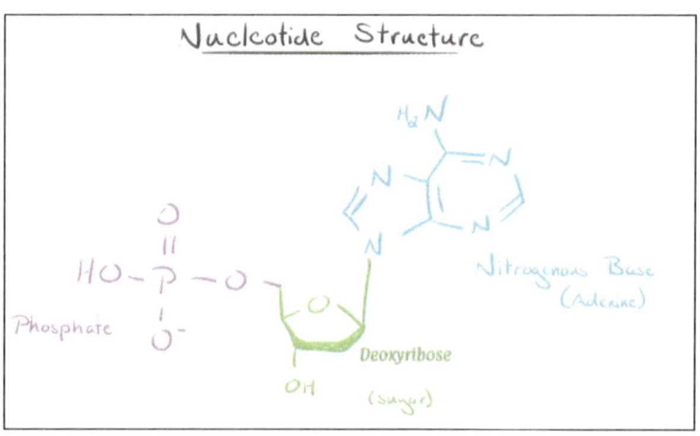

The phosphate group is a phosphorous atom with four oxygen atoms bonded to it. The phosphorous atom in phosphate has a marked tendency to bond to other oxygen atoms (for instance, the oxygen atom sticking off the deoxyribose sugar of another nucleotide).

Four Nucleotide Monomers are Distinguished by their Bases

Each type of nucleotide has a different nucleobase stuck to its deoxyribose sugar are:

- A nucleotide contains adenine,
- T nucleotide contains thymine,
- G nucleotide contains guanine,
- C nucleotide contains cytosine.

All four of these nucleobases are relatively complex molecules, with the unifying feature that they all tend to have multiple nitrogen atoms in their structures. For this reason, nucleobases are often also called nitrogenousbases.

1. Phosphodiester bonds in DNA polymers connect the 5' carbon of one nucleotide to the 3' carbon of another nucleotide.

The nucleotide monomers in a DNA polymer are connected by strong electromagnetic attractions called phosphodiester bonds. Phosphodiester bonds are part of a larger class of electromagnetic attractions between atoms that chemists refer to as covalent bonds.

In order to keep things organized, biochemists have developed a numbering system for talking about the molecular structure of nucleotides. These numbers are applied to the carbon atoms in the sugar, starting at the carbon immediately to the right of the oxygen in the deoxyribose ring, and continuing in a clockwise fashion: the numbers range from 1' ("one prime"), identifying the carbon immediately to the right of the oxygen) all the way to 5' ("five prime"), identifying the carbon that sticks off the fourth and final carbon in the deoxyribose ring.

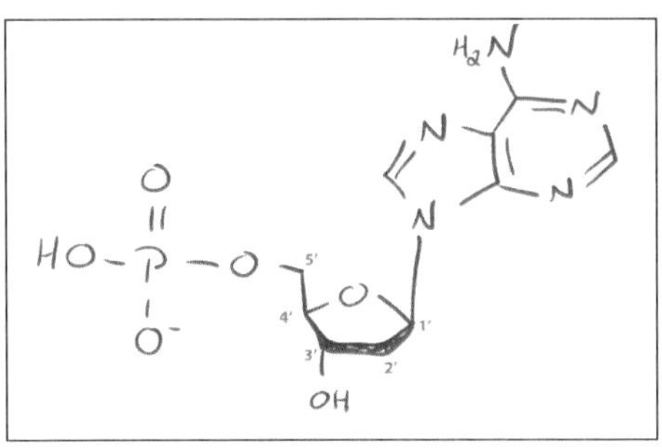

The phosphodiester bonds that join one DNA nucleotide to another always link the 3' carbon of the first nucleotide to the 5' carbon of the second nucleotide. This forms a covalent bond between the oxygen sticking off the 3' carbon of the first nucleotide, and the phosphorous atom in the phosphate group that sticks off the 5' carbon of the second nucleotide. These bonds are called 3'-5' phosphodiester bonds. Each time nucleotides are bound together, a water molecule is removed (or "lost") through a process called dehydration synthesis. Many molecules rely on dehydration synthesis to assist with forming polymers.

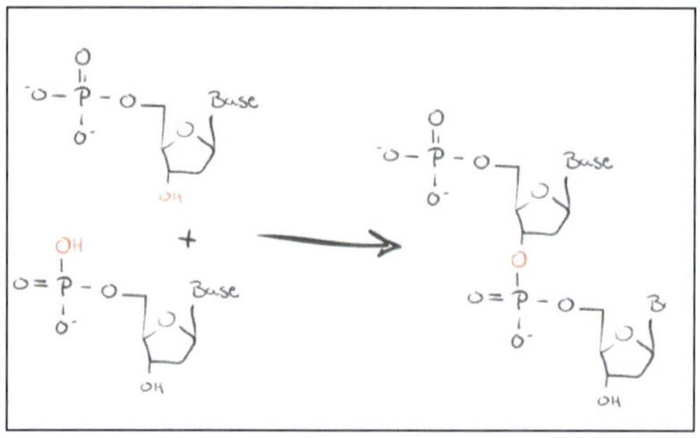

2. Chromosomes are made of two DNA polymers that stick together via non-covalent hydrogen bonds.

Chromosomal DNA consists of two DNA polymers that make up a 3-dimensional (3D) structure called a double helix. In a double helix structure, the strands of DNA run antiparallel, meaning the 5' end of one DNA strand is parallel with the 3' end of the other DNA strand.

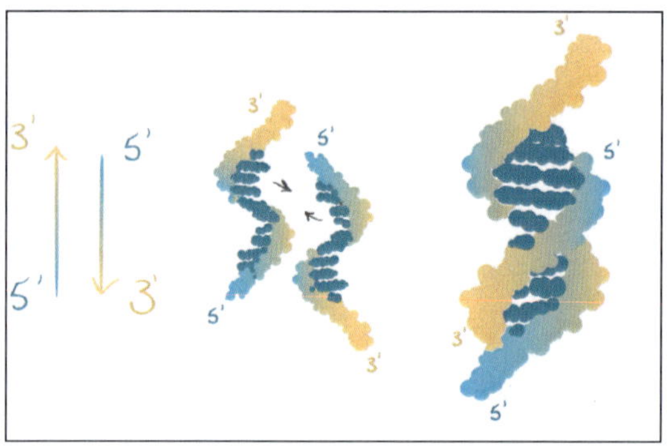

The nucleotides forming each DNA strand are connected by noncovalent bonds, called *hydrogen bonds*. Considered individually, hydrogen bonds are much weaker than a single covalent bond, such as a phosphodiester bond. But, there are so many of them that the two DNA polymers are very strongly connected to each other.

The hydrogen bonds that join DNA polymers happen between certain hydrogen atoms on one base (called hydrogen bond donors) and certain oxygen or nitrogen atoms on the base across from it (called hydrogen bond acceptors). Adenine ("A") and Thymine ("T") each have one donor and one acceptor, whereas Cytosine ("C") has one donor and two acceptors, and Guanine ("G") has one acceptor and two donors.

The A nucleotides are always hydrogen bonded to T nucleotides, and C nucleotides are always hydrogen bonded to G nucleotides. This selective binding is called *complementary base pairing*, and creates consistency in the nucleotide sequences of the two DNA polymers that join together to make a chromosome. This was first observed by

Erwin Chargaff, who developed methods for counting nucleotides in DNA samples, and found that the percent of A nucleotides always equaled the percent of T nucleotides, and the percent of G nucleotides always equaled the percent of C nucleotides (within a margin of error). Now, we know that complementary base pairing can be explained by reference to hydrogen bonding between the donors and acceptors on the bases of each nucleotide: A nucleotides and T nucleotides have a match (one donor and one acceptor each), and C nucleotides and G nucleotides have a match (the former has one donor and two acceptors, while the latter has one acceptor and two donors).

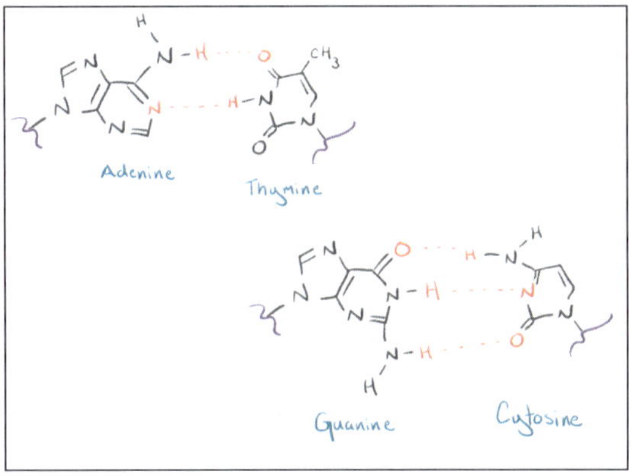

The Biological Function of DNA

1. DNA polymers direct the production of other polymers called proteins: A protein is one or more polymers of monomers called amino acids. Proteins are the workhorse molecules in your cells. They act as enzymes, structural support, hormones, and a whole host of other functional molecules. All traits derive from the interactions of proteins with each other and the surrounding environments.

2. A chromosome consists of smaller segments called genes: Chromosomes are very long structures consisting of two DNA polymers, joined together by hydrogen bonds connecting complementary base pairs. A chromosome is divided into segments of double-stranded DNA called genes.

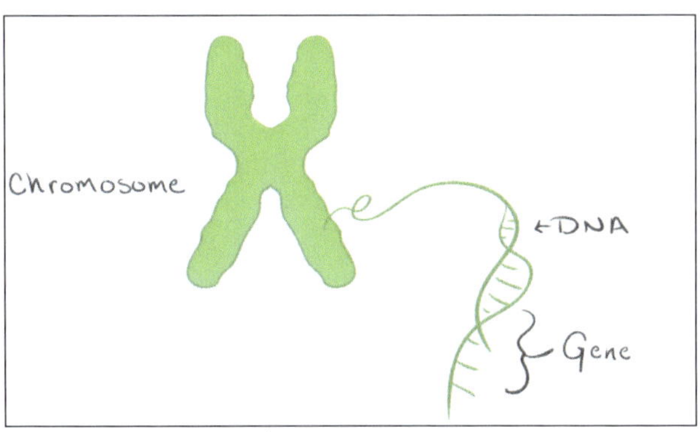

3. Each gene is further divided into three nucleotide subsegments called codons: A codon is a segment (or piece) of double stranded DNA that is three nucleotides long. A gene can be thought of as many three-nucleotide codons strung together.

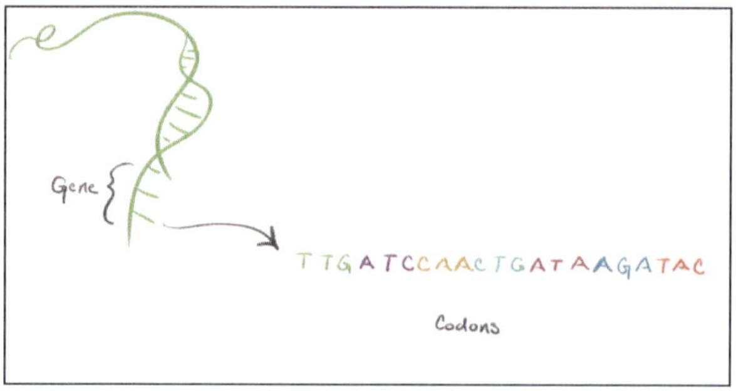

4. Understanding DNA structure and function: Earlier, we compared a DNA polymer to a sentence, and the nucleotide monomers that make up a polymer to the letters of the alphabet that are used to write sentences down. Now that we know what genes are, and what codons are, we can extend this analogy a bit further, and begin to get an insight into how DNA stores biological information.

If nucleotides are like letters, then codons are like words. Unlike English, where we use 26 letters to make words of all different lengths and meanings, your cells use the four DNA nucleotide monomers to make "words"—codons—of just one length: three nucleotides long. If you do the math, you'll see that this means that there are just 64 possible "words" in the DNA language—64 different ways of arranging the four DNA nucleotides into three-nucleotide-long combinations.

Just like in English, where each word is associated with a dictionary definition, the codons of the DNA language are each associated with specific amino acids. During translation on the ribosomes, each codon from the original DNA gene is matched with its corresponding amino acid (with the help of tRNA molecules). Just like a human reader puts the definitions of words together to arrive at the meaning of a sentence, a ribosome puts the amino acids referred to by each codon in a gene together, creating covalent bonds between them to make a protein.

A simplified example:

Imagine a basic sort of organism that only makes four proteins, each of which consists of four amino acid monomers. The traits of such an organism—how it eats, how it looks like, how it moves, etc.—are fully determined by the actions of these proteins.

The genes that specify how to make each of the four proteins are split across two chromosomes. This means that each chromosome consists of two genes. Since the proteins specified by the genes all have four amino acid monomers, each gene must have four codons. And, since a codon always consists of three nucleotides, each gene contains 12 nucleotide monomers, and, therefore, each chromosome is 24 nucleotides long.

DNA Replication

DNA is the genetic material that defines every cell. Before a cell duplicates and is divided into new daughter cells through either mitosis or meiosis, biomolecules and organelles must be copied to be distributed among the cells. DNA, found within the nucleus, must be replicated in order to ensure that each new cell receives the correct number of chromosomes. The process of DNA duplication is called DNA replication. Replication follows several steps that involve multiple proteins called replication enzymes and RNA. In eukaryotic cells, such as animal cells and plant cells, DNA replication occurs in the S phase of interphase during the cell cycle. The process of DNA replication is vital for cell growth, repair, and reproduction in organisms.

DNA Analysis

DNA analysis is the name given to the interpretation of genetic sequences, and can be used for a wide variety of purposes. It can be used to identify a species, but can also differentiate individuals within a species. Unsurprisingly, the DNA sequences of two different species vary more than those of two individuals from the same species. That said, a significant amount of DNA can still be shared between different species.

Perhaps the most recognisable examples of DNA analysis are the black and white images that contain bands (they look a bit similar to barcodes). Each of these bands represents a different fragment of DNA, and together they can act as a sort of 'genetic fingerprint' that can be used to compare different samples. Using this technique, a DNA sample from a crime scene can be quickly and easily matched to a suspect's DNA, or a biological relationship can be proved or disproved between a person and their alleged father.

DNA Samples

The process of DNA analysis starts with DNA being extracted and purified from a biological sample. DNA can be found in several types of samples and extracted using a range different techniques – the technique chosen often depends on the size of the sample and the amount of DNA likely to be present.

In the vast majority of cases, when taking a home DNA test, you'll be asked to provide a saliva sample or a cheek swab – a cheek swab is painless and takes just a few minutes. A few tests, most notably prenatal DNA tests, may require you to provide a small blood sample. Depending on the type of sample required, after you order the test, you'll receive a kit in the post that contains the appropriate sample collection equipment, with instructions for securely returning it to the lab.

Types of DNA Analysis

Regardless of the type of sample being analysed, once it has reached the lab and the DNA has been extracted, there are several methods of analysis that can be applied.

Restriction Fragment Length Polymorphism

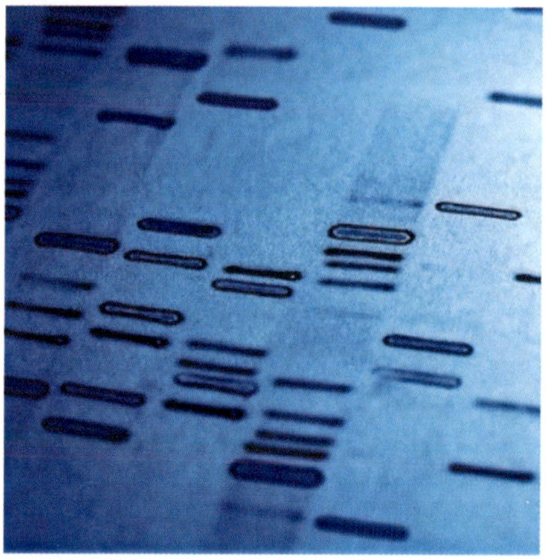

RFLP is one of the oldest types of DNA analysis and it produces the black and white images that we mentioned earlier. Put simply, this technique involves cutting up DNA by targeting specific sequences. The sequences are then cut into strands of differing lengths, and separated by length using a special gel. This produces the black and white 'genetic fingerprint' image, with the longest stands showing as bands at one end and the shortest at the other. By comparing two images you can compare the lengths of the strands and easily see whether the sequences are the same or different. This analysis usually needs a sample that contains a relatively high quantity of clean and contamination-free DNA. It also takes a lot of time, due to the numerous steps involved in the process and has therefore largely been replaced by newer, quicker methods.

Short Tandem Repeat Analysis

One such technique is the analysis of Short Tandem Repeats (STRs). STRs are sections of DNA, between one and five nucleotides, that are repeated several times at specific points in our DNA sequence. Compared to RFLP, STR analysis can be done with DNA samples of much lower quality, which means that even trace amounts of DNA can be analysed.

The analysis itself involves examining the number of times that certain STR sequences are repeated. These repeats occur in the same locations in everyone's DNA, but the amount of times that each STR repeats will differ between individuals. Those who are biologically related will possess STRs that are repeated the same or a similar number of times, and so this technique is the most commonly used when testing to prove or disprove biological relationships.

Single Nucleotide Polymorphism Analysis

Single Nucleotide Polymorphism (SNP) analysis is the most recently developed DNA analysis technique. SNPs are also known as genetic variants and are points in our DNA where there are single nucleotide differences between individuals. These are examined by running a DNA sample across a specialised computer chip. These chips are designed to detect up to one million SNPs in your DNA but will usually report on around 100,000. Analysis of these genetic variants is the method usually used to determine someone's genetic predisposition to disease, and is increasingly common in ancestry DNA testing.

Y DNA Analysis

The previous three techniques that we've talked about usually involve analysis of the DNA that's inherited from both parents, known as autosomal DNA. However, these DNA analysis techniques can also be applied to Y DNA, that only males possess and which is exclusively passed from father to son.

This technique can be used to help solve sexual assault cases in which there are several male suspects. Investigators use Y DNAanalysis to match samples taken from the victim with samples of suspects, to accurately find out who was involved. As the name indicates, the technique analyses multiple Y chromosome genetic markers, usually using STR or SNP analysis.

Y DNA analysis can also be utilised for tracing ancestral connections between males. It can provide information about 'deep' paternal ancestry, going back hundreds of thousands of years, as well as ancestral migratory paths.

Mitochondrial DNA (mtDNA) Analysis

In a similar way to Y DNA analysis, mitochondrial DNA analysis is concerned with the genetic information that we exclusively inherit from our mothers. However, unlike Y DNA which is only passed down from father to son, mitochondrial DNA is passed from mother to both son and daughter. It can be used to find out about your 'deep' maternal ancestry.

Although most of our DNA is found in the nuclei of our cells, mitochondrial DNA is found in mitochondria – structures in our cells that are separate from the nuclei. Interestingly, mitochondrial DNA testing is helpful for solving so-called 'cold cases', where the biological sample collected from a crime scene has degraded over time. This is because there are several mitochondria in each of our cells, instead of just one nucleus. Therefore, even if there is no nuclear DNA left, there is a chance that mitochondrial DNA remains intact and it can be analysed.

Uses of DNA Analysis

Paternity and other Relationship Tests

DNA can be analysed to find out if a particular individual is biologically related to another person, for example, by proving or disproving the biological link between an alleged father and his child.

In the case of paternity testing, multiple alleged fathers can be tested if there are doubts. Paternity testing can also be conducted before a child is born, by taking a blood sample from the mother, isolating the baby's DNA, and comparing it to that of the alleged father.

Ancestry

Human populations have been studied for centuries, but DNA analysis has provided a new tool with which to do this. Scientists can analyse samples taken from living individuals and archaeological remains across the globe, to find out how far back human populations go and to explore different ethnicities. For example, scientists used mitochondrial DNA analysis to conclude that all humans are blood-related to one woman, a woman who lived hundreds of thousands of years ago nicknamed 'Mitochondrial Eve'. Research such as this has been and continues to be used to develop consumer ancestry DNA kits that allow people to explore their lineage and their ethnic mix.

Identification

DNA analysis is used to identify human remains in criminal investigations and for archaeological research. However, DNA analysis is also used when those in high-risk jobs or environments (e.g. oil rig workers or military personnel) wish to obtain their DNA profile. DNA profiles can be used to identify someone in the event of a fatal accident as it's much more resilient than traditional forms of identification.

The US army now uses DNA profiles to supplement the traditional 'dog tag', and every new military recruit is required to provide saliva and blood samples. These can then be used to identify those killed while discharging duties. Outside the military however, DNA is rarely used for this purpose, as fingerprints and dental records are still preferred.

Predicting Disease

DNA can also be analysed to determine genetic links to diseases. Researchers study groups of people that suffer with a condition (e.g. Alzheimer's Disease) to try to establish the genetic factors that lead to an increased risk of suffering with the illness. This has led to the development of online DNA tests that can identify your 'genetic predisposition' to a number of conditions.

Genetic predisposition tests are understandably more strictly regulated than other types of DNA analysis, sometimes requiring a doctor's approval. It's also important to point out that these types of test don't diagnose a disease, but indicate whether you're at a higher risk of developing one. Rheumatoid arthritis, breast cancer and type 2 diabetes are just a few examples of conditions that a genetic test can reveal an increased or decreased genetic predisposition to.

Recombinant DNA

Recombinant DNA refers to the molecules of DNA from two different species that are inserted into a host organism to produce new genetic combinations that are of value to science, medicine, agriculture, and industry. Since the focus of all genetics is the gene, the fundamental goal of laboratory geneticists is to isolate, characterize, and manipulate genes. Although it is relatively easy to isolate a sample of DNA from a collection of cells, finding a specific gene within this DNA sample can be compared to finding a needle in a haystack. Consider the fact that each human cell contains approximately 2 metres (6 feet) of DNA. Therefore, a small tissue sample will contain many kilometres of DNA. However, recombinant DNA technology has made it possible to isolate one gene or any other segment of DNA, enabling researchers to determine its nucleotide sequence, study its transcripts, mutate it in highly specific ways, and reinsert the modified sequence into a living organism.

In Vitro Mutagenesis

Another use of cloned DNA is in vitro mutagenesis in which a mutation is produced in a segment of cloned DNA. The DNA is then inserted into a cell or organism, and the effects of the mutation are studied. Mutations are useful to geneticists in enabling them to investigate the components of any biological process. However, traditional mutational analysis relied on the occurrence of random spontaneous mutations—a hit-or-miss method in which it was impossible to predict the precise type or position of the mutations obtained. In vitro mutagenesis, however, allows specific mutations to be tailored for type and for position within the gene. A cloned gene is treated in the test tube (in vitro) to obtain the specific mutation desired, and then this fragment is reintroduced into the living cell, where it replaces the resident gene.

One method of in vitro mutagenesis is oligonucleotide-directed mutagenesis. A specific point in a sequenced gene is pinpointed for mutation. An oligonucleotide, a short stretch of synthetic DNA of the desired sequence, is made chemically. For example, the oligonucleotide might have adenine in one specific location instead of guanine. This oligonucleotide is hybridized to the complementary strand of the cloned gene; it will hybridize despite the one base pair mismatch. Various enzymes are added to allow the oligonucleotide to prime the synthesis of a complete strand within the vector. When the vector is introduced into a bacterial cell and replicates, the mutated strand will act as a template for a complementary strand that will also be mutant, and thus a fully mutant molecule is obtained. This fully mutant cloned molecule is then reintroduced into the donor organism, and the mutant DNA replaces the resident gene.

Another version of in vitro mutagenesis is gene disruption, or gene knockout. Here, the resident functional gene is replaced by a completely nonfunctional copy. The advantage of this technique over random mutagenesis is that specific genes can be knocked out at will, leaving all other genes untouched by the mutagenic procedure.

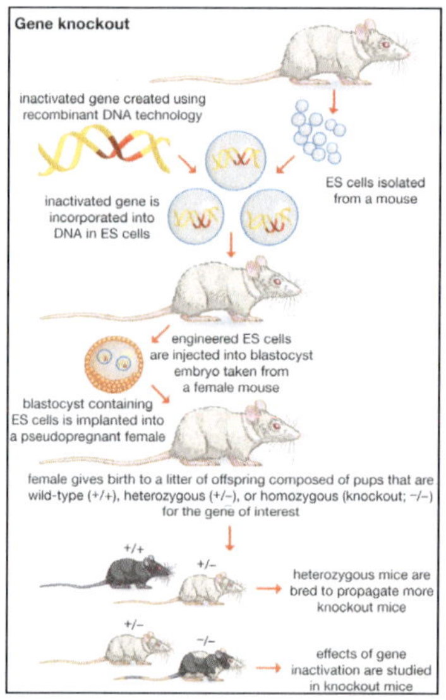

In gene knockout a functional gene is replaced by an inactivated gene that is created using recombinant DNA technology. When a gene is "knocked out," the resulting mutant phenotype (observable characteristics) often reveals the gene's biological function.

Genetically Modified Organisms

The ability to obtain specific DNA clones using recombinant DNA technology has made it possible to add the DNA of one organism to the genome of another. The added gene is called a transgene. The transgene inserts itself into a chromosome and is passed to the progeny as a new component of the genome. The resulting organism carrying the transgene is called a transgenic organism or a genetically modified organism (GMO). In this way, a "designer organism" is made that contains some specific change required for an experiment in basic genetics or for improvement of some commercial strain.

Several transgenic plants have been produced. Genes for toxins that kill insects have been introduced in several species, including corn and cotton. Bacterial genes that confer resistance to herbicides also have been introduced into crop plants. Other plant transgenes aim at improving the nutritional value of the plant.

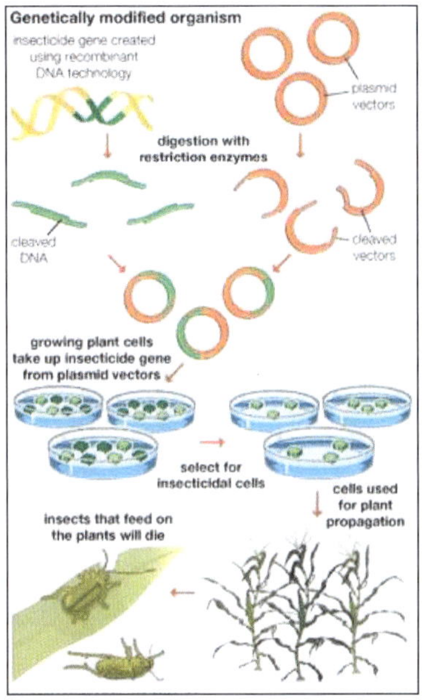

Genetically modified organisms are produced using scientific methods that include recombinant DNA technology.

Genetically modified corn (maize).

Gene Therapy

Gene therapy is the introduction of a normal gene into an individual's genome in order to repair a mutation that causes a genetic disease. When a normal gene is inserted into a mutant nucleus, it most likely will integrate into a chromosomal site different from the defective allele; although this may repair the mutation, a new mutation may result if the normal gene integrates into another functional gene. If the normal gene replaces the mutant allele, there is a chance that the transformed cells will proliferate and produce enough normal gene product for the entire body to be restored to the undiseased phenotype. So far, human gene therapy has been attempted only on somatic (body) cells for diseases such as cancer and severe combined immunodeficiency syndrome (SCIDS). Somatic cells cured by gene therapy may reverse the symptoms of disease in the treated individual, but the modification is not passed on to the next generation. Germinal gene therapy aims to place corrected cells inside the germ line (e.g., cells of the ovary or testis). If this is achieved, these cells will undergo meiosis and provide a normal gametic contribution to the next generation. Germinal gene therapy has been achieved experimentally in animals but not in humans.

Reverse Genetics

Recombinant DNA technology has made possible a type of genetics called reverse genetics. Traditionally, genetic research starts with a mutant phenotype, and, by Mendelian crossing analysis, a researcher is able to attribute the phenotype to a specific gene. Reverse genetics travels in precisely the opposite direction. Researchers begin with a gene of unknown function and use molecular analysis to determine its phenotype. One important tool in reverse genetics is gene knockout. By mutating the cloned gene of unknown function and using it to replace the resident copy or copies, the resultant mutant phenotype will show which biological function this gene normally controls.

Diagnostics

Recombinant DNA technology has led to powerful diagnostic procedures useful in both medicine and forensics. In medicine these diagnostic procedures are used in counseling prospective parents as to the likelihood of having a child with a particular disease, and they are also used in the prenatal prediction of genetic disease in the fetus. Researchers look for specific DNA fragments that are located in close proximity to the gene that causes the disease of concern. These fragments, called restriction fragment length polymorphisms (RFLPs), often serve as effective "genetic markers." In forensics, DNA fragments called variable number tandem repeats (VNTRs), which are highly variable between individuals, are employed to produce what is called a "DNA fingerprint." A DNA fingerprint can be used to determine if blood or other body fluids left at the scene of a crime belongs to a suspect.

Genomics

The genetic analysis of entire genomes is called genomics. Such a broadscale analysis has been made possible by the development of recombinant DNA technology. In humans, knowledge of the entire genome sequence has facilitated searching for genes that produce hereditary diseases. It is also capable of revealing a set of proteins—produced at specific times, in specific tissues, or in specific diseases—that might be targets for therapeutic drugs. Genomics also allows the comparison of one genome with another, leading to insights into possible evolutionary relationships between organisms.

Genomics has two subdivisions: structural genomics and functional genomics. Structural genomics is based on the complete nucleotide sequence of a genome. Each member of a library of clones is physically manipulated by robots and sequenced by automatic sequencing machines, enabling a very high throughput of DNA. The resulting sequences are then assembled by a computer into a complete sequence for every chromosome. The complete DNA sequence is scanned by computer to find the positions of open reading frames (ORFs), or prospective genes. The sequences are then compared to the sequences of known genes from other organisms, and possible functions are assigned.

Functional genomics attempts to understand function at the broadest level (the genomic level). In one approach, gene functions of as many ORFs as possible are assigned as above in an attempt to obtain a full set of proteins encoded by the genome (called a proteome). The proteome broadly defines all the cellular functions used by the organism. Function in relation to specific developmental stages also is assessed by trying to identify the "transcriptome," the set of mRNA transcripts made at specific developmental stages. The practical approach utilizes microarrays—glass plates the size of a microscope slide imprinted with tens of thousands of ordered DNA samples, each representing one gene (either a clone or a synthesized segment). The mRNA preparation under test is labeled with a fluorescent dye, and the microarray is bathed in this mRNA. Fluorescent spots appear on the array indicating which mRNAs were present, thus defining the transcriptome.

Protein Manufacture

Recombinant DNA procedures have been used to convert bacteria into "factories" for the synthesis of foreign proteins. This technique is useful not only for preparing large amounts of protein for basic research but also for producing valuable proteins for medical use. For example, the genes for human proteins such as growth hormone, insulin, and blood-clotting factor can be commercially manufactured. Another approach to producing proteins via recombinant DNA technology is to introduce the desired gene into the genome of an animal, engineered in such a way that the protein is secreted in the animal's milk, facilitating harvesting.

DNA Sequencing

DNA sequencing is the process of determining the sequence of nucleotide bases (As, Ts, Cs, and Gs) in a piece of DNA. Today, with the right equipment and materials, sequencing a short piece of DNA is relatively straightforward.

Sequencing an entire genome (all of an organism's DNA) remains a complex task. It requires breaking the DNA of the genome into many smaller pieces, sequencing the pieces, and assembling the sequences into a single long "consensus." However, thanks to new methods that have been developed over the past two decades, genome sequencing is now much faster and less expensive than it was during the Human Genome Project.

We'll take a look at methods used for DNA sequencing. We'll focus on one well-established method, Sanger sequencing, but we'll also discuss new ("next-generation") methods that have reduced the cost and accelerated the speed of large-scale sequencing.

Sanger Sequencing: The Chain Termination Method

Regions of DNA up to about 900 base pairs in length are routinely sequenced using a method called Sanger sequencing or the chain termination method. Sanger sequencing was developed by the British biochemist Fred Sanger and his colleagues.

In the Human Genome Project, Sanger sequencing was used to determine the sequences of many relatively small fragments of human DNA. (These fragments weren't necessarily 900 bp or less, but researchers were able to "walk" along each fragment using multiple rounds of Sanger sequencing.) The fragments were aligned based on overlapping portions to assemble the sequences of larger regions of DNA and, eventually, entire chromosomes.

Although genomes are now typically sequenced using other methods that are faster and less expensive, Sanger sequencing is still in wide use for the sequencing of individual pieces of DNA, such as fragments used in DNA cloning or generated through polymerase chain reaction (PCR).

Ingredients for Sanger Sequencing

Sanger sequencing involves making many copies of a target DNA region. Its ingredients are similar to those needed for DNA replication in an organism, or for polymerase chain reaction (PCR), which copies DNA *in vitro*. They include:

- A DNA polymerase enzyme.
- A primer, which is a short piece of single-stranded DNA that binds to the template DNA and acts as a "starter" for the polymerase.
- The four DNA nucleotides (dATP, dTTP, dCTP, dGTP).
- The template DNA to be sequenced.

However, a Sanger sequencing reaction also contains a unique ingredient:

- Dideoxy, or chain-terminating, versions of all four nucleotides (ddATP, ddTTP, ddCTP, ddGTP), each labeled with a different color of dye.

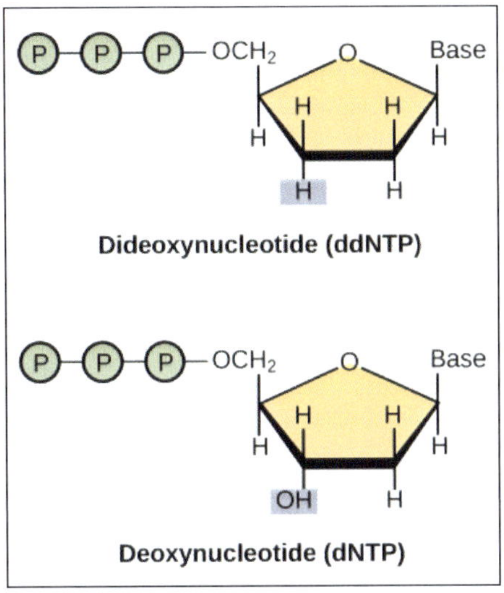

Whole-genome sequencing.

Dideoxy nucleotides are similar to regular, or deoxy, nucleotides, but with one key difference: they lack a hydroxyl group on the 3' carbon of the sugar ring. In a regular nucleotide, the 3' hydroxyl group acts as a "hook," allowing a new nucleotide to be added to an existing chain.

Once a dideoxy nucleotide has been added to the chain, there is no hydroxyl available and no further nucleotides can be added. The chain ends with the dideoxy nucleotide, which is marked with a particular color of dye depending on the base (A, T, C or G) that it carries.

Method of Sanger Sequencing

The DNA sample to be sequenced is combined in a tube with primer, DNA polymerase, and DNA nucleotides (dATP, dTTP, dGTP, and dCTP). The four dye-labeled, chain-terminating dideoxy nucleotides are added as well, but in much smaller amounts than the ordinary nucleotides.

The mixture is first heated to denature the template DNA (separate the strands), then cooled so that the primer can bind to the single-stranded template. Once the primer has bound, the temperature is raised again, allowing DNA polymerase to synthesize new DNA starting from the primer. DNA polymerase will continue adding nucleotides to the chain until it happens to add a dideoxy nucleotide instead of a normal one. At that point, no further nucleotides can be added, so the strand will end with the dideoxy nucleotide.

This process is repeated in a number of cycles. By the time the cycling is complete, it's virtually guaranteed that a dideoxy nucleotide will have been incorporated at every single position of the target DNA in at least one reaction. That is, the tube will contain fragments of different lengths, ending at each of the nucleotide positions in the original

DNA. The ends of the fragments will be labeled with dyes that indicate their final nucleotide.

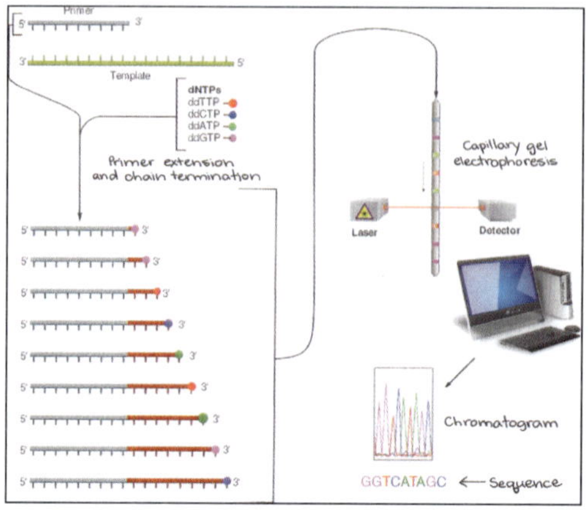

Sanger sequencing.

After the reaction is done, the fragments are run through a long, thin tube containing a gel matrix in a process called capillary gel electrophoresis. Short fragments move quickly through the pores of the gel, while long fragments move more slowly. As each fragment crosses the "finish line" at the end of the tube, it's illuminated by a laser, allowing the attached dye to be detected.

The smallest fragment (ending just one nucleotide after the primer) crosses the finish line first, followed by the next-smallest fragment (ending two nucleotides after the primer), and so forth. Thus, from the colors of dyes registered one after another on the detector, the sequence of the original piece of DNA can be built up one nucleotide at a time. The data recorded by the detector consist of a series of peaks in fluorescence intensity, as shown in the chromatogram above. The DNA sequence is read from the peaks in the chromatogram.

Uses and Limitations

Sanger sequencing gives high-quality sequence for relatively long stretches of DNA (up to about 900 base pairs). It's typically used to sequence individual pieces of DNA, such as bacterial plasmids or DNA copied in PCR.

However, Sanger sequencing is expensive and inefficient for larger-scale projects, such as the sequencing of an entire genome or metagenome (the "collective genome" of a microbial community). For tasks such as these, new, large-scale sequencing techniques are faster and less expensive.

Next-generation Sequencing

The name may sound like Star Trek, but that's really what it's called. The most recent set of DNA sequencing technologies are collectively referred to as next-generation sequencing.

There are a variety of next-generation sequencing techniques that use different technologies. However, most share a common set of features that distinguish them from Sanger sequencing:

- Highly parallel: Many sequencing reactions take place at the same time.
- Micro scale: Reactions are tiny and many can be done at once on a chip.
- Fast: Because reactions are done in parallel, results are ready much faster.
- Low-cost: Sequencing a genome is cheaper than with Sanger sequencing.
- Shorter length: reads typically range from 50 -700 nucleotides in length.

Conceptually, next-generation sequencing is kind of like running a very large number of tiny Sanger sequencing reactions in parallel. Thanks to this parallelization and small scale, large quantities of DNA can be sequenced much more quickly and cheaply with next-generation methods than with Sanger sequencing.

There are about 3 billion base pairs of DNA in your genome, all of which must be accurately copied when any one of your trillions of cells divides. The basic mechanisms of DNA replication are similar across organisms.

Basic Idea

DNA replication is semiconservative, meaning that each strand in the DNA double helix acts as a template for the synthesis of a new, complementary strand.

This process takes us from one starting molecule to two "daughter" molecules, with each newly formed double helix containing one new and one old strand.

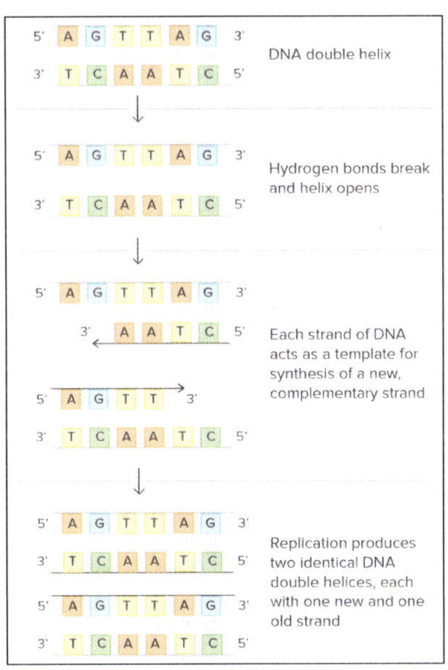

In a sense, that's all there is to DNA replication! But what's actually most interesting about this process is how it's carried out in a cell.

Cells need to copy their DNA very quickly, and with very few errors (or risk problem such as cancer). To do so, they use a variety of enzymes and proteins, which work together to make sure DNA replication is performed smoothly and accurately.

DNA Polymerase

One of the key molecules in DNA replication is the enzyme DNA polymerase. DNA polymerases are responsible for synthesizing DNA: they add nucleotides one by one to the growing DNA chain, incorporating only those that are complementary to the template.

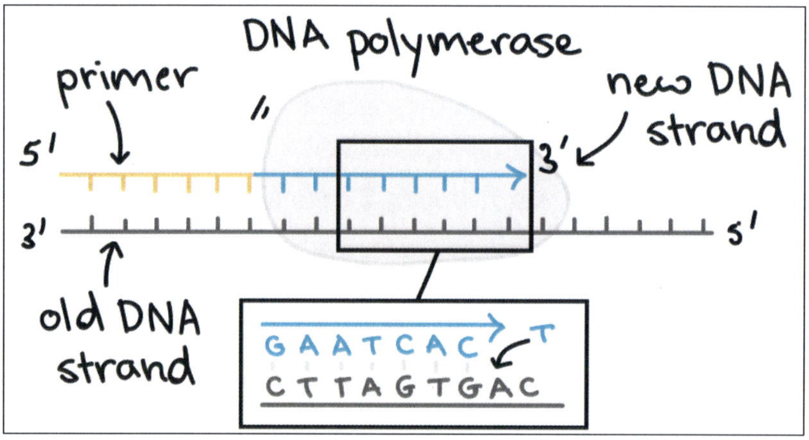

Here are some key features of DNA polymerases:

- They always need a template.
- They can only add nucleotides to the 3' end of a DNA strand.
- They can't start making a DNA chain from scratch, but require a pre-existing chain or short stretch of nucleotides called a primer.
- They proofread, or check their work, removing the vast majority of "wrong" nucleotides that are accidentally added to the chain.

The addition of nucleotides requires energy. This energy comes from the nucleotides themselves, which have three phosphates attached to them (much like the energy-carrying molecule ATP). When the bond between phosphates is broken, the energy released is used to form a bond between the incoming nucleotide and the growing chain.

In prokaryotes such as E. coli, there are two main DNA polymerases involved in DNA replication: DNA pol III (the major DNA-maker), and DNA pol I, which plays a crucial supporting role.

Starting DNA Replication

Replication always starts at specific locations on the DNA, which are called origins of replication and are recognized by their sequence.

E. coli, like most bacteria, has a single origin of replication on its chromosome. The origin is about 245 base pairs long and has mostly A/T base pairs (which are held together by fewer hydrogen bonds than G/C base pairs), making the DNA strands easier to separate.

Specialized proteins recognize the origin, bind to this site, and open up the DNA. As the DNA opens, two Y-shaped structures called replication forks are formed, together making up what's called a replication bubble. The replication forks will move in opposite directions as replication proceeds.

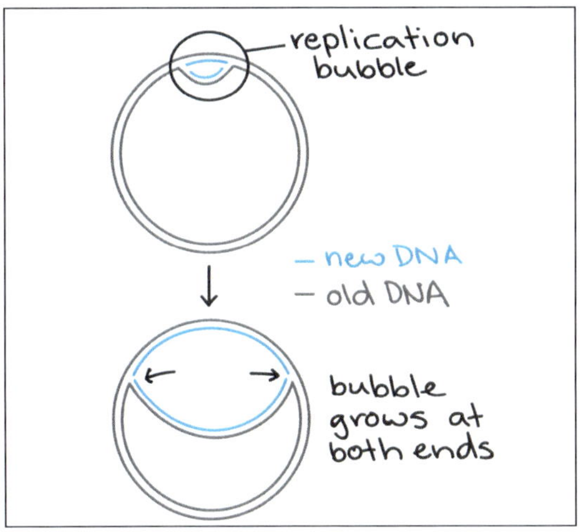

How does replication actually get going at the forks? Helicase is the first replication enzyme to load on at the origin of replication end superscript. Helicase's job is to move the replication forks forward by "unwinding" the DNA (breaking the hydrogen bonds between the nitrogenous base pairs).

Proteins called single-strand binding proteins coat the separated strands of DNA near the replication fork, keeping them from coming back together into a double helix.

Primers and Primase

DNA polymerases can only add nucleotides to the 3' end of an existing DNA strand. (They use the free -OH group found at the 3' end as a "hook," adding a nucleotide to this group in the polymerization reaction.) How, then, does DNA polymerase add the first nucleotide at a new replication fork?

Alone, it can't! The problem is solved with the help of an enzyme called primase. Primase makes an RNA primer, or short stretch of nucleic acid complementary to the template, that provides a 3' end for DNA polymerase to work on. A typical primer is about five to ten nucleotides long. The primer *primes* DNA synthesis, i.e., gets it started.

Once the RNA primer is in place, DNA polymerase "extends" it, adding nucleotides one by one to make a new DNA strand that's complementary to the template strand.

Leading and Lagging Strands

In *E. coli*, the DNA polymerase that handles most of the synthesis is DNA polymerase III. There are two molecules of DNA polymerase III at a replication fork, each of them hard at work on one of the two new DNA strands.

DNA polymerases can only make DNA in the 5' to 3' direction, and this poses a problem during replication. A DNA double helix is always anti-parallel; in other words, one strand runs in the 5' to 3' direction, while the other runs in the 3' to 5' direction. This makes it necessary for the two new strands, which are also antiparallel to their templates, to be made in slightly different ways.

One new strand, which runs 5' to 3' towards the replication fork, is the easy one. This strand is made continuously, because the DNA polymerase is moving in the same direction as the replication fork. This continuously synthesized strand is called the leading strand.

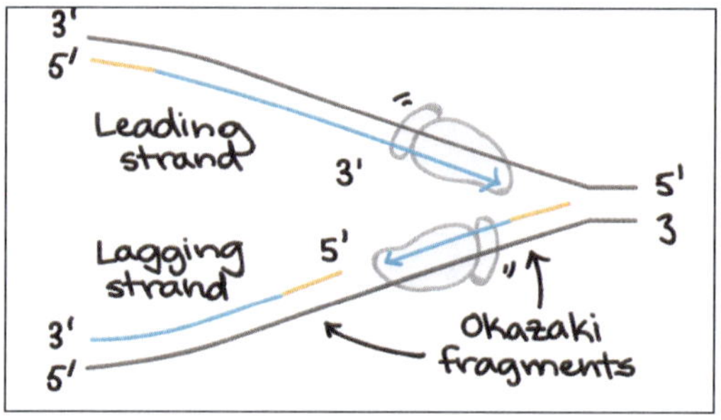

The other new strand, which runs 5' to 3' away from the fork, is trickier. This strand is made in fragments because, as the fork moves forward, the DNA polymerase (which is moving away from the fork) must come off and reattach on the newly exposed DNA. This tricky strand, which is made in fragments, is called the lagging strand.

The small fragments are called Okazaki fragments, named for the Japanese scientist who discovered them. The leading strand can be extended from one primer alone, whereas the lagging strand needs a new primer for each of the short Okazaki fragments.

Maintenance and Cleanup

Some other proteins and enzymes, in addition the main ones above, are needed to keep DNA replication running smoothly. One is a protein called the sliding clamp, which holds DNA polymerase III molecules in place as they synthesize DNA. The sliding clamp is a ring-shaped protein and keeps the DNA polymerase of the lagging strand from floating off when it re-starts at a new Okazaki fragment.

Topoisomerase also plays an important maintenance role during DNA replication. This enzyme prevents the DNA double helix ahead of the replication fork from getting too tightly wound as the DNA is opened up. It acts by making temporary nicks in the helix to release the tension, then sealing the nicks to avoid permanent damage.

Finally, there is a little cleanup work to do if we want DNA that doesn't contain any RNA or gaps. The RNA primers are removed and replaced by DNA through the activity of DNA polymerase I, the other polymerase involved in replication. The nicks that remain after the primers are replaced get sealed by the enzyme DNA ligase.

DNA Cloning

DNA cloning is the process of making multiple, identical copies of a particular piece of DNA. In a typical DNA cloning procedure, the gene or other DNA fragment of interest (perhaps a gene for a medically important human protein) is first inserted into a circular piece of DNA called a plasmid. The insertion is done using enzymes that "cut and paste" DNA, and it produces a molecule of recombinant DNA, or DNA assembled out of fragments from multiple sources.

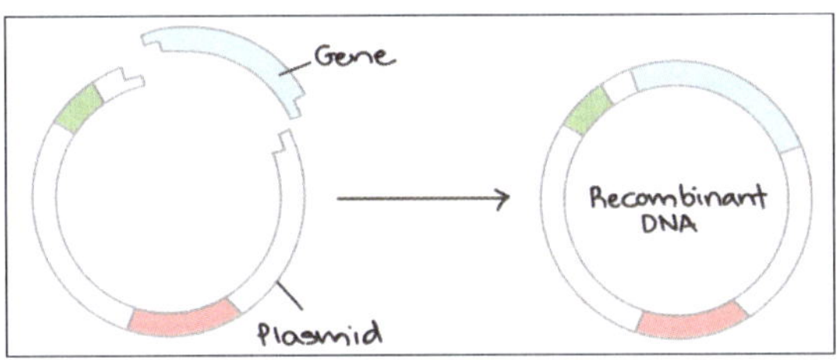

Next, the recombinant plasmid is introduced into bacteria. Bacteria carrying the plasmid are selected and grown up. As they reproduce, they replicate the plasmid and pass it on to their offspring, making copies of the DNA it contains.

What is the point of making many copies of a DNA sequence in a plasmid? In some cases, we need lots of DNA copies to conduct experiments or build new plasmids. In other cases, the piece of DNA encodes a useful protein, and the bacteria are used as "factories" to make the protein. For instance, the human insulin gene is expressed in E. coli bacteria to make insulin used by diabetics.

Steps of DNA Cloning

DNA cloning is used for many purposes. As an example, DNA cloning can be used to synthesize a protein (such as human insulin) in bacteria. The basic steps are:

1. Cut open the plasmid and "paste" in the gene. This process relies on restriction enzymes (which cut DNA) and DNA ligase (which joins DNA).

2. Insert the plasmid into bacteria. Use antibiotic selection to identify the bacteria that took up the plasmid.

3. Grow up lots of plasmid-carrying bacteria and use them as "factories" to make the protein. Harvest the protein from the bacteria and purify it.

Cutting and Pasting DNA

How can pieces of DNA from different sources be joined together? A common method uses two types of enzymes: restriction enzymes and DNA ligase.

A restriction enzyme is a DNA-cutting enzyme that recognizes a specific target sequence and cuts DNA into two pieces at or near that site. Many restriction enzymes produce cut ends with short, single-stranded overhangs. If two molecules have matching overhangs, they can base-pair and stick together. However, they won't combine to form an unbroken DNA molecule until they are joined by DNA ligase, which seals gaps in the DNA backbone.

Our goal in cloning is to insert a target gene (e.g., for human insulin) into a plasmid. Using a carefully chosen restriction enzyme, we digest:

- The plasmid, which has a single cut site.

- The target gene fragment, which has a cut site near each end.

Then, we combine the fragments with DNA ligase, which links them to make a recombinant plasmid containing the gene.

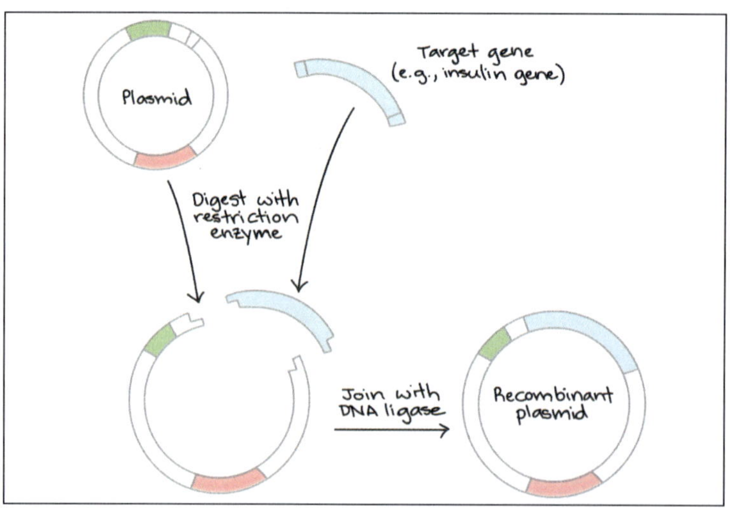

Bacterial Transformation and Selection

Plasmids and other DNA can be introduced into bacteria, such as the harmless E. coli used in labs, in a process called transformation. During transformation, specially prepared bacterial cells are given a shock (such as high temperature) that encourages them to take up foreign DNA.

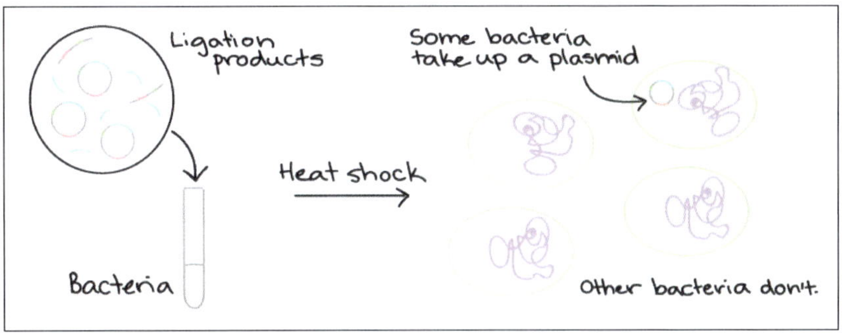

A plasmid typically contains an antibiotic resistance gene, which allows bacteria to survive in the presence of a specific antibiotic. Thus, bacteria that took up the plasmid can be selected on nutrient plates containing the antibiotic. Bacteria without a plasmid will die, while bacteria carrying a plasmid can live and reproduce. Each surviving bacterium will give rise to a small, dot-like group, or colony, of identical bacteria that all carry the same plasmid.

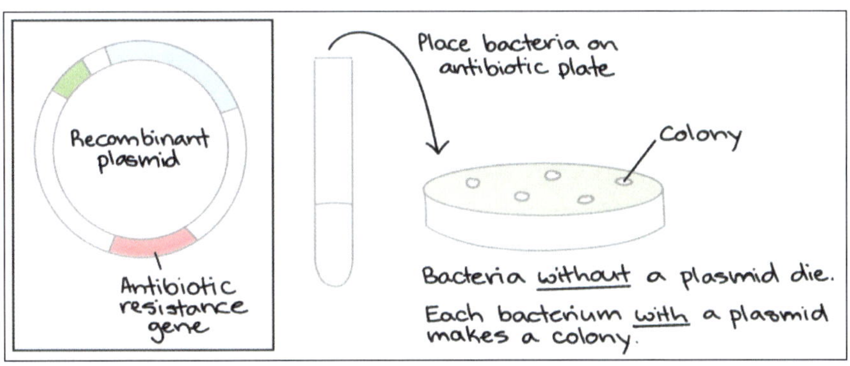

Not all colonies will necessarily contain the right plasmid. That's because, during a ligation, DNA fragments don't always get "pasted" in exactly the way we intend. Instead, we must collect DNA from several colonies and see whether each one contain the right plasmid. Methods like restriction enzyme digestion and PCR are commonly used to check the plasmids.

Protein Production

Once we have found a bacterial colony with the right plasmid, we can grow a large culture of plasmid-bearing bacteria. Then, we give the bacteria a chemical signal that instructs them to make the target protein.

The bacteria serve as miniature "factories," churning out large amounts of protein. For instance, if our plasmid contained the human insulin gene, the bacteria would start transcribing the gene and translating the mRNA to produce many molecules of human insulin protein.

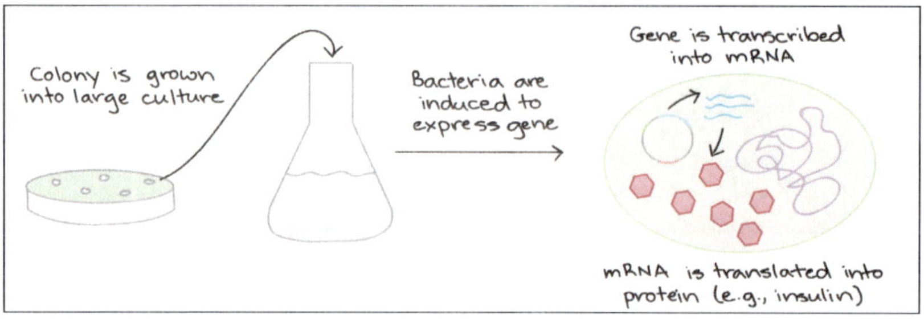

Once the protein has been produced, the bacterial cells can be split open to release it. There are many other proteins and macromolecules floating around in bacteria besides the target protein (e.g., insulin). Because of this, the target protein must be purified, or separated from the other contents of the cells by biochemical techniques. The purified protein can be used for experiments or, in the case of insulin, administered to patients.

Uses of DNA Cloning

DNA molecules built through cloning techniques are used for many purposes in molecular biology. A short list of examples includes:

- Biopharmaceuticals: DNA cloning can be used to make human proteins with biomedical applications, such as the insulin mentioned above. Other examples of recombinant proteins include human growth hormone, which is given to patients who are unable to synthesize the hormone, and tissue plasminogen activator (tPA), which is used to treat strokes and prevent blood clots. Recombinant proteins like these are often made in bacteria.

- Gene therapy: In some genetic disorders, patients lack the functional form of a particular gene. Gene therapy attempts to provide a normal copy of the gene to the cells of a patient's body. For example, DNA cloning was used to build plasmids containing a normal version of the gene that's nonfunctional in cystic fibrosis. When the plasmids were delivered to the lungs of cystic fibrosis patients, lung function deteriorated less quickly.

- Gene analysis: In basic research labs, biologists often use DNA cloning to build artificial, recombinant versions of genes that help them understand how normal genes in an organism function.

These are just a few examples of how DNA cloning is used in biology today. DNA cloning is a very common technique that is used in a huge variety of molecular biology applications.

RNA

Structurally speaking, ribonucleic acid (RNA), is quite similar to DNA. However, whereas DNA molecules are typically long and double stranded, RNA molecules are much shorter and are typically single stranded. RNA molecules perform a variety of roles in the cell but are mainly involved in the process of protein synthesis (translation) and its regulation.

RNA Structure

RNA is typically single stranded and is made of ribonucleotides that are linked by phosphodiester bonds. A ribonucleotide in the RNA chain contains ribose (the pentose sugar), one of the four nitrogenous bases (A, U, G, and C), and a phosphate group. The subtle structural difference between the sugars gives DNA added stability, making DNA more suitable for storage of genetic information, whereas the relative instability of RNA makes it more suitable for its more short-term functions. The RNA-specific pyrimidine uracil forms a complementary base pair with adenine and is used instead of the thymine used in DNA. Even though RNA is single stranded, most types of RNA molecules show extensive intramolecular base pairing between complementary sequences within the RNA strand, creating a predictable three-dimensional structure essential for their function.

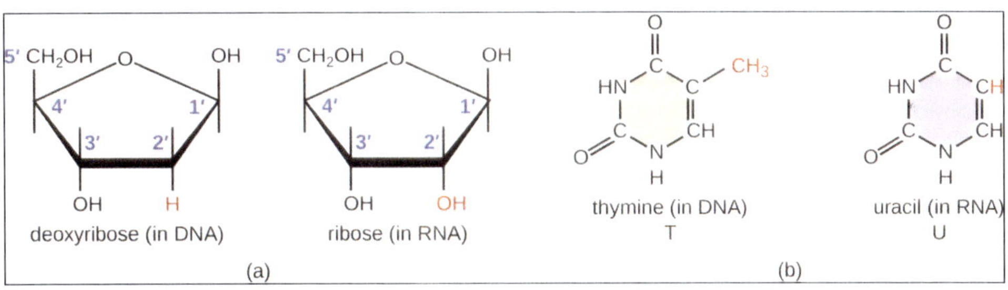

(a) Ribonucleotides contain the pentose sugar ribose instead of the deoxyribose found in deoxyribonucleotides. (b) RNA contains the pyrimidine uracil in place of thymine found in DNA.

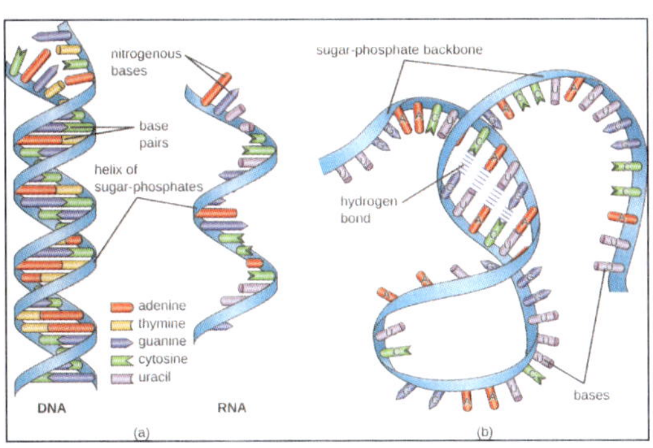

a) DNA is typically double stranded, whereas RNA is typically single stranded. (b) Although it is single stranded, RNA can fold upon itself, with the folds stabilized by short areas of complementary base pairing within the molecule, forming a three-dimensional structure.

Functions of RNA in Protein Synthesis

Cells access the information stored in DNA by creating RNA to direct the synthesis of proteins through the process of translation. Proteins within a cell have many functions, including building cellular structures and serving as enzyme catalysts for cellular chemical reactions that give cells their specific characteristics. The three main types of RNA directly involved in protein synthesis are messenger RNA (mRNA), ribosomal RNA (rRNA), and transfer RNA (tRNA).

In 1961, French scientists François Jacob and Jacques Monod hypothesized the existence of an intermediary between DNA and its protein products, which they called messenger RNA. Evidence supporting their hypothesis was gathered soon afterwards showing that information from DNA is transmitted to the ribosome for protein synthesis using mRNA. If DNA serves as the complete library of cellular information, mRNA serves as a photocopy of specific information needed at a particular point in time that serves as the instructions to make a protein.

The mRNA carries the message from the DNA, which controls all of the cellular activities in a cell. If a cell requires a certain protein to be synthesized, the gene for this product is "turned on" and the mRNA is synthesized through the process of transcription. The mRNA then interacts with ribosomes and other cellular machinery to direct the synthesis of the protein it encodes during the process of translation. mRNA is relatively unstable and short-lived in the cell, especially in prokaryotic cells, ensuring that proteins are only made when needed.

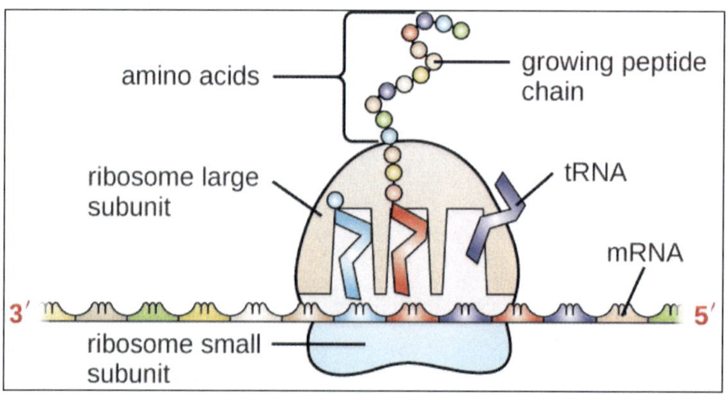

A generalized illustration of how mRNA and tRNA are used in protein synthesis within a cell.

rRNA and tRNA are stable types of RNA. In prokaryotes and eukaryotes, tRNA and rRNA are encoded in the DNA, then copied into long RNA molecules that are cut to release smaller fragments containing the individual mature RNA species. In eukaryotes, synthesis, cutting, and assembly of rRNA into ribosomes takes place in the nucleolus region of the nucleus, but these activities occur in the cytoplasm of prokaryotes. Neither of these types of RNA carries instructions to direct the synthesis of a polypeptide, but they play other important roles in protein synthesis.

Ribosomes are composed of rRNA and protein. As its name suggests, rRNA is a major constituent of ribosomes, composing up to about 60% of the ribosome by mass and providing the location where the mRNA binds. The rRNA ensures the proper alignment of the mRNA, tRNA, and the ribosomes; the rRNA of the ribosome also has an enzymatic activity (peptidyl transferase) and catalyzes the formation of the peptide bonds between two aligned amino acids during protein synthesis. Although rRNA had long been thought to serve primarily a structural role, its catalytic role within the ribosome was proven in 2000.

Transfer RNA is the third main type of RNA and one of the smallest, usually only 70–90 nucleotides long. It carries the correct amino acid to the site of protein synthesis in the ribosome. It is the base pairing between the tRNA and mRNA that allows for the correct amino acid to be inserted in the polypeptide chain being synthesized. Any mutations in the tRNA or rRNA can result in global problems for the cell because both are necessary for proper protein synthesis.

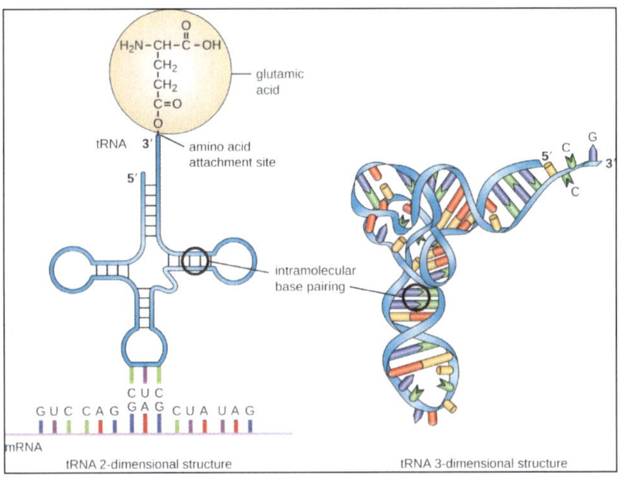

A tRNA molecule is a single-stranded molecule that exhibits significant intracellular base pairing, giving it its characteristic three-dimensional shape.

Table: Structure and Function of RNA.

	mRNA	**rRNA**	**tRNA**
Structure	Short, unstable, single-stranded RNA corresponding to a gene encoded within DNA.	Longer, stable RNA molecules composing 60% of ribosome's mass.	Short (70-90 nucleotides), stable RNA with extensive intramolecular base pairing; contains an amino acid binding site and an mRNA binding site.
Function	Serves as intermediary between DNA and protein; used by ribosome to direct synthesis of protein it encodes.	Ensures the proper alignment of mRNA, tRNA, and ribosome during protein synthesis; catalyzes peptide bond formation between amino acids.	Carries the correct amino acid to the site of protein synthesis in the ribosome.

RNA as Hereditary Information

Although RNA does not serve as the hereditary information in most cells, RNA does hold this function for many viruses that do not contain DNA. Thus, RNA clearly does have the additional capacity to serve as genetic information. Although RNA is typically single stranded within cells, there is significant diversity in viruses. Rhinoviruses, which cause the common cold; influenza viruses; and the Ebola virus are single-stranded RNA viruses. Rotaviruses, which cause severe gastroenteritis in children and other immunocompromised individuals, are examples of double-stranded RNA viruses. Because double-stranded RNA is uncommon in eukaryotic cells, its presence serves as an indicator of viral infection.

RNA Processing

The newly made RNA, also known as the primary transcript (the product of transcription is known as a transcript) is further processed before it is functional. Both prokaryotes and eukaryotes process their ribosomal and transfer RNAs.

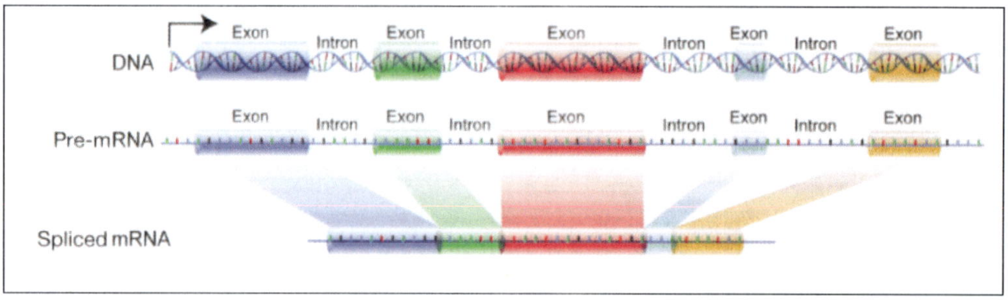

RNA splicing.

The major difference in RNA processing, however, between prokaryotes and eukaryotes, is in the processing of messenger RNAs. In eukaryotic cells, RNA synthesis, which occurs in the nucleus, is separated from the protein synthesis machinery, which is in the cytoplasm. In addition, eukaryotic genes have introns, noncoding regions that interrupt the gene's coding sequence. The mRNA copied from genes containing introns will also therefore have regions that interrupt the information in the gene. These regions must be removed before the mRNA is sent out of the nucleus to be used to direct protein synthesis. The process of removing the introns and rejoining the coding sections or exons, of the mRNA, is called splicing. Once the mRNA has been capped, spliced and had a polyA tail added, it is sent from the nucleus into the cytoplasm for translation.

The initial product of transcription of a protein coding gene is called the pre-mRNA (or primary transcript). After it has been processed and is ready to be exported from the nucleus, it is called the mature mRNA or processed mRNA.

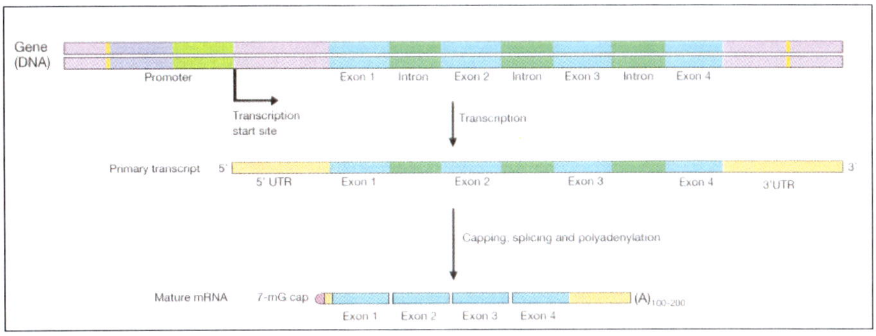

Steps in processing eukaroytic messenger RNAs.

In eukaryotic cells, pre-mRNAs undergo three main processing steps:

- Capping at the 5' end,
- Addition of a polyA tail at the 3' end, and
- Splicing to remove introns.

Capping

5' end Capping

Post-transcriptional processing of the 5' end of the RNA product of DNA transcription comes in the form of a process called the 5' cap. At the end of transcription, the 5' end of the RNA transcript contains a free triphosphate group since it was the first incorporated nucleotide in the chain. The capping process replaces the triphosphate group with another structure called the "cap". The cap is added by the enzyme guanyl transferase. This enzyme catalyzes the reaction between the 5' end of the RNA transcript and a guanine triphosphate (GTP) molecule.

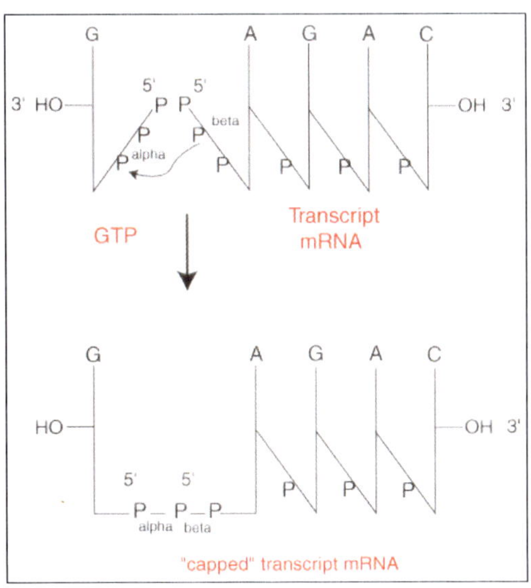

The 5' Capping Reaction.

The figure, simply illustrates the reaction between the 5' end of the RNA transcript and the GTP molecule. In the reaction, the beta phosphate of the RNA transcript displaces a pyrophosphate group at the 5' position of the GTP molecule. The cap is formed through a 5'-5' linkage between the two substrates such that the GTP molecule is oriented in the opposite direction as the other nucleotides in the RNA transcript chain.

Once in place, the cap plays a role in the ribosomal recognition of messenger RNA during translation into a protein. Prokaryotes do not have a similar cap because they use other signals for recognition by the ribosome.

Polyadenylation

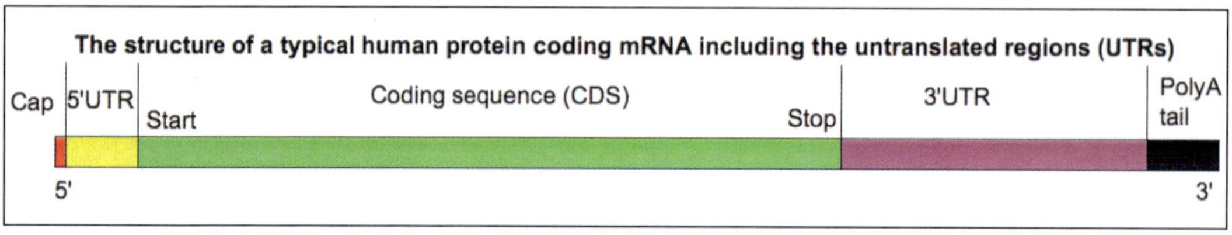

Typical structure of a mature eukaryotic mRNA.

Polyadenylation is the addition of a poly(A) tail to a messenger RNA. The poly(A) tail consists of multiple adenosine monophosphates; in other words, it is a stretch of RNA that has only adenine bases. In eukaryotes, polyadenylation is part of the process that produces mature messenger RNA (mRNA) for translation. It, therefore, forms part of the larger process of gene expression.

The process of polyadenylation begins as the transcription of a gene terminates. The 3'-most segment of the newly made pre-mRNA is first cleaved off by a set of proteins; these proteins then synthesize the poly(A) tail at the RNA's 3' end. In some genes these proteins add a poly(A) tail at one of several possible sites. Therefore, polyadenylation can produce more than one transcript from a single gene (alternative polyadenylation), similar to alternative splicing.

The poly(A) tail is important for the nuclear export, translation, and stability of mRNA. The tail is shortened over time, and, when it is short enough, the mRNA is enzymatically degraded. However, in a few cell types, mRNAs with short poly(A) tails are stored for later activation by re-polyadenylation in the cytosol. In contrast, when polyadenylation occurs in bacteria, it promotes RNA degradation. This is also sometimes the case for eukaryotic non-coding RNAs.

mRNA molecules in both prokaryotes and eukaryotes have polyadenylated 3'-ends, with the prokaryotic poly(A) tails generally shorter and less mRNA molecules polyadenylated.

Nuclear Polyadenylation

Function

In nuclear polyadenylation, a poly(A) tail is added to an RNA at the end of transcription. On mRNAs, the poly(A) tail protects the mRNA molecule from enzymatic degradation in the cytoplasm and aids in transcription termination, export of the mRNA from the nucleus, and translation. Almost all eukaryotic mRNAs are polyadenylated, with the exception of animal replication-dependent histone mRNAs. These are the only mRNAs in eukaryotes that lack a poly(A) tail, ending instead in a stem-loop structure followed by a purine-rich sequence, termed histone downstream element, that directs where the RNA is cut so that the 3' end of the histone mRNA is formed.

Many eukaryotic non-coding RNAs are always polyadenylated at the end of transcription. There are small RNAs where the poly(A) tail is seen only in intermediary forms and not in the mature RNA as the ends are removed during processing, the notable ones being microRNAs. But, for many long noncoding RNAs – a seemingly large group of regulatory RNAs that, for example, includes the RNA Xist, which mediates X chromosome inactivation – a poly(A) tail is part of the mature RNA.

Mechanism

The processive polyadenylation complex in the nucleus of eukaryotes works on products of RNA polymerase II, such as precursor mRNA. Here, a multi-protein complex cleaves the 3'-most part of a newly produced RNA and polyadenylates the end produced by this cleavage. The cleavage is catalysed by the enzyme CPSF and occurs 10–30 nucleotides downstream of its binding site. This site often has the polyadenylation signal sequence AAUAAA on the RNA, but variants of it that bind more weakly to CPSF exist. Two other proteins add specificity to the binding to an RNA: CstF and CFI. CstF binds to a GU-rich region further downstream of CPSF's site. CFI recognises a third site on the RNA (a set of UGUAA sequences in mammals) and can recruit CPSF even if the AAUAAA sequence is missing. The polyadenylation signal – the sequence motif recognised by the RNA cleavage complex – varies between groups of eukaryotes. Most human polyadenylation sites contain the AAUAAA sequence, but this sequence is less common in plants and fungi.

The RNA is typically cleaved before transcription termination, as CstF also binds to RNA polymerase II. Through a poorly understood mechanism, it signals for RNA polymerase II to slip off of the transcript. Cleavage also involves the protein CFII, though it is unknown how. The cleavage site associated with a polyadenylation signal can vary up to some 50 nucleotides.

When the RNA is cleaved, polyadenylation starts, catalysed by polyadenylate polymerase. Polyadenylate polymerase builds the poly(A) tail by adding adenosine monophosphate units from adenosine triphosphate to the RNA, cleaving off pyrophosphate.

Another protein, PAB2, binds to the new, short poly(A) tail and increases the affinity of polyadenylate polymerase for the RNA. When the poly(A) tail is approximately 250 nucleotides long the enzyme can no longer bind to CPSF and polyadenylation stops, thus determining the length of the poly(A) tail. CPSF is in contact with RNA polymerase II, allowing it to signal the polymerase to terminate transcription. When RNA polymerase II reaches a "termination sequence" ($^{5'}$TTTATT$^{3'}$ on the DNA template and $^{5'}$AAUAAA$^{3'}$) on the primary transcript), the end of transcription is signaled. The polyadenylation machinery is also physically linked to the spliceosome, a complex that removes introns from RNAs.

Downstream Effects

The poly(A) tail acts as the binding site for poly(A)-binding protein. Poly(A)-binding protein promotes export from the nucleus and translation, and inhibits degradation. This protein binds to the poly(A) tail prior to mRNA export from the nucleus and in yeast also recruits poly(A) nuclease, an enzyme that shortens the poly(A) tail and allows the export of the mRNA. Poly(A)-binding protein is exported to the cytoplasm with the RNA. mRNAs that are not exported are degraded by the exosome. Poly(A)-binding protein also can bind to, and thus recruit, several proteins that affect translation, one of these is initiation factor-4G, which in turn recruits the 40S ribosomal subunit. However, a poly(A) tail is not required for the translation of all mRNAs. Further, poly(A) tailing (oligo-adenylation) can determine the fate of RNA molecules that are usually not poly(A)-tailed (such as (small) non-coding (sn)RNAs etc.) and thereby induce their RNA decay.

Deadenylation

In eukaryotic somatic cells, the poly(A) tail of most mRNAs in the cytoplasm gradually get shorter, and mRNAs with shorter poly(A) tail are translated less and degraded sooner. However, it can take many hours before an mRNA is degraded. This deadenylation and degradation process can be accelerated by microRNAs complementary to the 3' untranslated region of an mRNA. In immature egg cells, mRNAs with shortened poly(A) tails are not degraded, but are instead stored without being translated. They are then activated by cytoplasmic polyadenylation after fertilisation, during egg activation.

In animals, poly(A) ribonuclease (PARN) can bind to the 5' cap and remove nucleotides from the poly(A) tail. The level of access to the 5' cap and poly(A) tail is important in controlling how soon the mRNA is degraded. PARN deadenylates less if the RNA is bound by the initiation factors 4E (at the 5' cap) and 4G (at the poly(A) tail), which is why translation reduces deadenylation. The rate of deadenylation may also be regulated by RNA-binding proteins. Once the poly(A) tail is removed, the decapping complex removes the 5' cap, leading to a degradation of the RNA. Several other enzymes that seem to be involved in deadenylation have been identified in yeast.

Alternative Polyadenylation

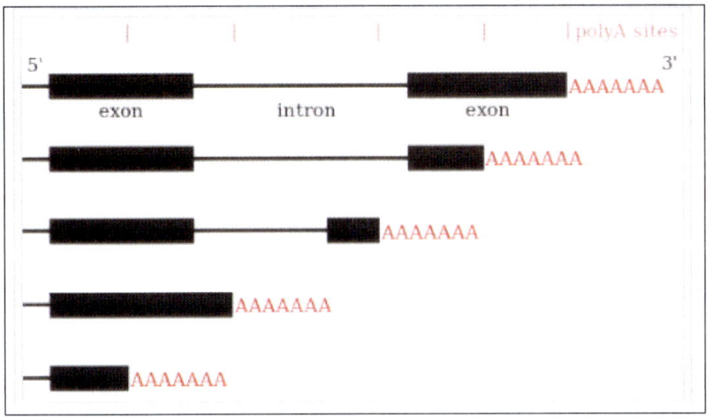

Results of using different polyadenylation sites on the same gene.

Many protein-coding genes have more than one polyadenylation site, so a gene can code for several mRNAs that differ in their 3' end. Since alternative polyadenylation changes the length of the 3' untranslated region, it can change which binding sites for microRNAs the 3' untranslated region contains. MicroRNAs tend to repress translation and promote degradation of the mRNAs they bind to, although there are examples of microRNAs that stabilise transcripts. Alternative polyadenylation can also shorten the coding region, thus making the mRNA code for a different protein, but this is much less common than just shortening the 3' untranslated region.

The choice of poly(A) site can be influenced by extracellular stimuli and depends on the expression of the proteins that take part in polyadenylation. For example, the expression of CstF-64, a subunit of cleavage stimulatory factor (CstF), increases in macrophages in response to lipopolysaccharides (a group of bacterial compounds that trigger an immune response). This results in the selection of weak poly(A) sites and thus shorter transcripts. This removes regulatory elements in the 3' untranslated regions of mRNAs for defense-related products like lysozyme and TNF-α. These mRNAs then have longer half-lives and produce more of these proteins. RNA-binding proteins other than those in the polyadenylation machinery can also affect whether a polyadenyation site is used, as can DNA methylation near the polyadenylation signal.

Cytoplasmic Polyadenylation

There is polyadenylation in the cytosol of some animal cell types, namely in the germ line, during early embryogenesis and in post-synaptic sites of nerve cells. This lengthens the poly(A) tail of an mRNA with a shortened poly(A) tail, so that the mRNA will be translated. These shortened poly(A) tails are often less than 20 nucleotides, and are lengthened to around 80–150 nucleotides.

In the early mouse embryo, cytoplasmic polyadenylation of maternal RNAs from the egg cell allows the cell to survive and grow even though transcription does not start until the middle of the 2-cell stage (4-cell stage in human). In the brain, cytoplasmic polyadenylation is active during learning and could play a role in long-term potentiation, which is the strengthening of the signal transmission from a nerve cell to another in response to nerve impulses and is important for learning and memory formation.

Cytoplasmic polyadenylation requires the RNA-binding proteins CPSF and CPEB, and can involve other RNA-binding proteins like Pumilio. Depending on the cell type, the polymerase can be the same type of polyadenylate polymerase (PAP) that is used in the nuclear process, or the cytoplasmic polymerase GLD-2.

Tagging for Degradation in Eukaryotes

For many non-coding RNAs, including tRNA, rRNA, snRNA, and snoRNA, polyadenylation is a way of marking the RNA for degradation, at least in yeast. This polyadenylation is done in the nucleus by the TRAMP complex, which maintains a tail that is around 4 nucleotides long to the 3' end. The RNA is then degraded by the exosome. Poly(A) tails have also been found on human rRNA fragments, both the form of homopolymeric (A only) and heterpolymeric (mostly A) tails.

Prokaryotes and Organelles

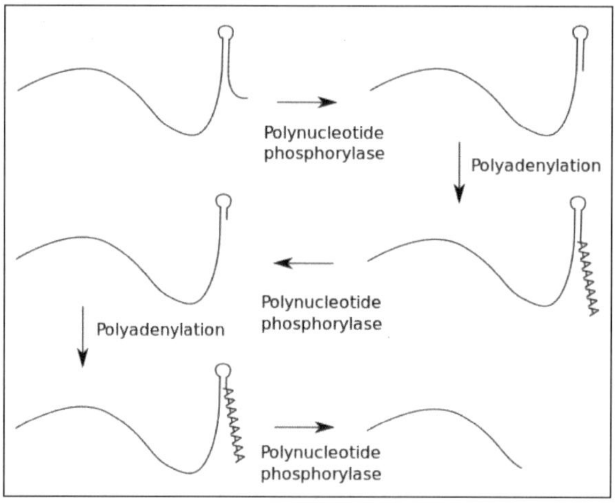

Polyadenylation in bacteria helps polynucleotide phosphorylase degrade past secondary structure.

In many bacteria, both mRNAs and non-coding RNAs can be polyadenylated. This poly(A) tail promotes degradation by the degradosome, which contains two RNA-degrading enzymes: polynucleotide phosphorylase and RNase E. Polynucleotide phosphorylase binds to the 3' end of RNAs and the 3' extension provided by the poly(A) tail allows it to bind to the RNAs whose secondary structure would otherwise block the 3' end. Successive rounds of polyadenylation and degradation of the 3' end by polynucleotide phosphorylase allows the degradosome to overcome these secondary structures. The poly(A) tail can also recruit RNases that cut the RNA in two. These bacterial poly(A) tails are about 30 nucleotides long.

In as different groups as animals and trypanosomes, the mitochondria contain both stabilising and destabilising poly(A) tails. Destabilising polyadenylation targets both mRNA and noncoding RNAs. The poly(A) tails are 43 nucleotides long on average. The stabilising ones start at the stop codon, and without them the stop codon (UAA) is not complete as the genome only encodes the U or UA part. Plant mitochondria have only destabilising polyadenylation, and yeast mitochondria have no polyadenylation at all.

While many bacteria and mitochondria have polyadenylate polymerases, they also have another type of polyadenylation, performed by polynucleotide phosphorylase itself. This enzyme is found in bacteria, mitochondria, plastids and as a constituent of the archaeal exosome (in those archaea that have an exosome). It can synthesise a 3' extension where the vast majority of the bases are adenines. Like in bacteria, polyadenylation by polynucleotide phosphorylase promotes degradation of the RNA in plastids and likely also archaea.

RNA Splicing

The third big RNA processing event that happens in your cells is RNA splicing. In RNA splicing, specific parts of the pre-mRNA, called introns are recognized and removed by a protein-and-RNA complex called the spliceosome. Introns can be viewed as "junk" sequences that must be cut out so the "good parts version" of the RNA molecule can be assembled.

The pieces of the RNA that are not chopped out are called exons. The exons are pasted together by the spliceosome to make the final, mature mRNA that is shipped out of the nucleus.

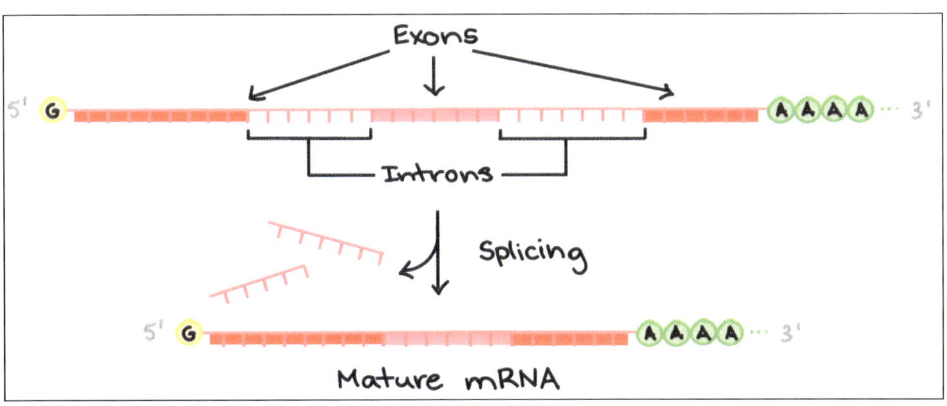

A key point here is that it's only the exons of a gene that encode a protein. Not only do the introns not carry information to build a protein, they actually have to be removed in order for the mRNA to encode a protein with the right sequence. If the spliceosome fails to remove an intron, an mRNA with extra "junk" in it will be made, and a wrong protein will get produced during translation.

Alternative Splicing

Splicing allow for a process called alternative splicing, in which more than one mRNA can be made from the same gene. Through alternative splicing, we (and other eukaryotes) can sneakily encode more different proteins than we have genes in our DNA.

In alternative splicing, one pre-mRNA may be spliced in either of two (or sometimes many more than two!) different ways. For example, in the diagram below, the same pre-mRNA can be spliced in three different ways, depending on which exons are kept. This results in three different mature mRNAs, each of which translates into a protein with a different structure.

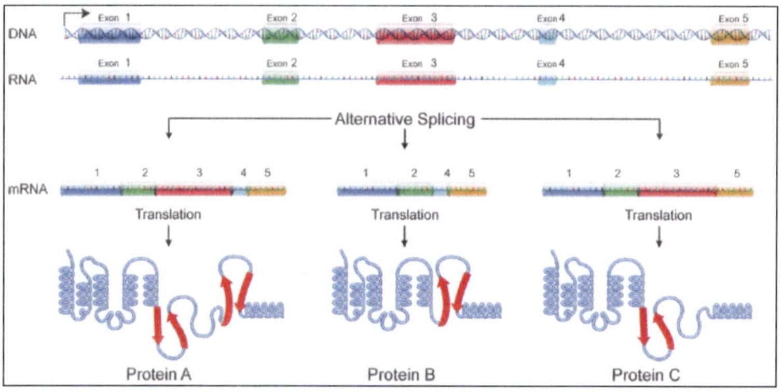

DNA, alternative splicing," by the National Human Genome Research Institute (public domain).

RNA Interference

RNA interference (RNAi) is a regulatory system occurring within eukaryotic cells (cells with a clearly defined nucleus) that controls the activity of genes. RNAi functions specifically to silence, or deactivate, genes.

The ability of interfering RNA to silence genes was discovered in the 1990s by American scientists Andrew Z. Fire and Craig C. Mello, who shared the 2006 Nobel Prize for Physiology or Medicine for their work. Fire and Mello successfully inhibited the expression of specific genes by introducing short double-stranded RNA (dsRNA) segments into the cells of nematodes (Caenorhabditis elegans). The dsRNA segments underwent enzymatic processing that enabled them to attach to molecules of messenger RNA (mRNA) possessing complementary nucleotide sequences. The attachment of the two RNAs inhibited the translation of the mRNA molecules into proteins.

RNAi in Nature

Although Fire and Mello's work involved the experimental introduction of interfering RNA into cells, gene silencing by RNAi is a natural genetic mechanism in eukaryotes that takes place following transcription (the synthesis of mRNA from DNA). Special microRNA (miRNA) segments, each of which is approximately 20 nucleotides in length, are encoded by the genomes of eukaryotic organisms. Each miRNA is produced from a precursor transcript (pre-miRNA). After the pre-miRNA migrates from the nucleus into the cytoplasm, it is cleaved into a mature miRNA by an enzyme known as DICER. The mature miRNA molecule then binds to an RNA-induced silencing complex (RISC), which contains multiple proteins, including a ribonuclease enzyme. The miRNA nucleotide sequence directs the protein complex to bind to a complementary sequence of mRNA. Once bound to the mRNA, the miRNA-RISC complex then enzymatically cleaves targeted sites on the mRNA molecule, thereby inhibiting the translation of the gene into a protein, which effectively silences the gene.

RNAi plays an important role not only in regulating genes but also in mediating cellular defense against infection by RNA viruses, including influenza viruses and rhabdoviruses, a group that contains the causative agent of rabies. In fact, a number of plants and animals have evolved antiviral RNAi genes that encode short segments of RNA molecules with sequences that are complementary to viral sequences. This complementarity enables interfering RNA produced by the cell to bind to and inactivate specific RNA viruses.

RNAi also is an innate mechanism by which cells can suppress the activity of transposons, or "jumping genes." Certain types of transposable elements are able to produce mobile copies of themselves, which subsequently are inserted into various regions of the genome, giving rise to repetitive sequences of DNA. These insertions generally are of little concern. However, some insertions lead to increased or decreased gene activity and can give rise to disease in humans. For example, certain types of cancer and Duchenne muscular dystrophy, a hereditary muscle-wasting disorder, are associated with insertions of transposons.

RNAi in Research and Medicine

The discovery that genes can be silenced by segments of dsRNA that are introduced into cells in tissue culture revolutionized the study of gene function. Gene silencing by dsRNA makes use of the naturally occurring cell machinery that is involved in the processing of miRNA in eukaryotic cells. For example, each dsRNA is cleaved into small pieces by the DICER enzyme. These pieces are called short interfering RNAs (siRNAs) and are about 20 to 25 nucleotides in length. Similar to miRNA, siRNA binds to RISC and cleaves targeted sequences of mRNA.

There are different types of synthetic dsRNAs that can be employed to disrupt gene function. Commonly used molecules include siRNA, which bypasses DICER cleavage, and small hairpin RNA (shRNA), which actually is one RNA strand containing two unique siRNA segments that is folded into a double strand, with the adjacent nucleotides joined through heating (annealing) rather than complementary base pairing. This creates a structure that resembles a hairpin because it has a tight loop at one end. Inside a cell, the shRNA is cleaved into its two component siRNAs by DICER.

RNAi is an exceptionally powerful research tool. Synthetic dsRNAs are designed to prevent the expression of specific genes, thereby enabling geneticists to manipulate the activity of genes in order to better understand their functions. In addition, abnormally overactive genes contribute to certain human diseases, and silencing this activity using RNAi has become an important area of investigation. RNAi is being explored as a form of treatment for a variety of diseases, including macular degeneration, hepatitis, AIDS, Huntington disease, and cancer.

In macular degeneration, RNA sequences that block the production of a protein called vascular endothelial growth factor (VEGF) in cells of the retina can inhibit the excess growth of retinal blood vessels, which leak and lead to vision loss. RNAi treatments for macular degeneration involve the injection of "naked RNA" into the eye. The term naked RNA is used to distinguish this approach from those that employ viral vectors to introduce dsRNA into diseased cells. Interfering RNAs incorporated into vectors are being studied for their effectiveness in slowing tumour growth. For example, mRNA transcripts of genes known to be overactive in certain forms of cancer serve as useful targets for RNAi-based treatments, which can silence overactive genes and slow disease progression.

Factors such as ensuring that interfering RNAs reach the cells and that the viral vectors themselves do not give rise to dangerous side effects have complicated the development of RNAi therapies. Furthermore, sequence similarities between genes can result in the binding of dsRNAs to otherwise properly functioning genes. This can result in the silencing of healthy genes vital to normal cell function. Nevertheless, the technique remains promising for applications in medicine.

References

- Dna-structure-and-function, dna, biomolecules: khanacademy.org, Retrieved 3 April, 2019
- Dna-replication-3981005: thoughtco.com, Retrieved 30 July, 2019
- What-is-dna-analysis: dnatestingchoice.com, Retrieved 19 January, 2019
- Recombinant-DNA-technology, science: britannica.com, Retrieved 9 April, 2019
- Dna-sequencing, dna-sequencing-pcr-electrophoresis, biotech-dna-technology, biology, science: khanacademy.org, Retrieved 17 February, 2019
- Mattick JS, Makunin IV (April 2006). "Non-coding RNA". Human Molecular Genetics. 15 Spec No 1 (90001): R17–29. doi:10.1093/hmg/ddl046. PMID 16651366
- Molecular-mechanism-of-dna-replication, dna-replication, dna-as-the-genetic-material, biology, science: khanacademy.org, Retrieved 26 August, 2019
- Dna-cloning, biotech-dna-technology, biology, science: khanacademy.org, Retrieved 6 March, 2019
- Structure-and-function-of-rna, microbiology: lumenlearning.com, Retrieved 12 May, 2019
- Lehninger AL, Nelson DL, Cox MM, eds. (1993). Principles of biochemistry (2nd ed.). New York: Worth. ISBN 978-0-87901-500-8
- RNA-interference, science: britannica.com, Retrieved 20 June, 2019

Chapter 4

Gene Expression

The process by which information from a gene is used in the synthesis of a functional gene product is known as gene expression. The various steps which are a part of gene expression are transcription, translation and post-translational modification of protein. These different components of gene expression as well as gene mutation have been thoroughly discussed in this chapter.

Gene expression is the process by which the genetic code - the nucleotide sequence - of a gene is used to direct protein synthesis and produce the structures of the cell. Genes that code for amino acid sequences are known as 'structural genes'.

The process of gene expression involves two main stages:

- Transcription: the production of messenger RNA (mRNA) by the enzyme RNA polymerase, and the processing of the resulting mRNA molecule.
- Translation: the use of mRNA to direct protein synthesis, and the subsequent post-translational processing of the protein molecule.

Some genes are responsible for the production of other forms of RNA that play a role in translation, including transfer RNA (tRNA) and ribosomal RNA (rRNA).

A structural gene involves a number of different components:

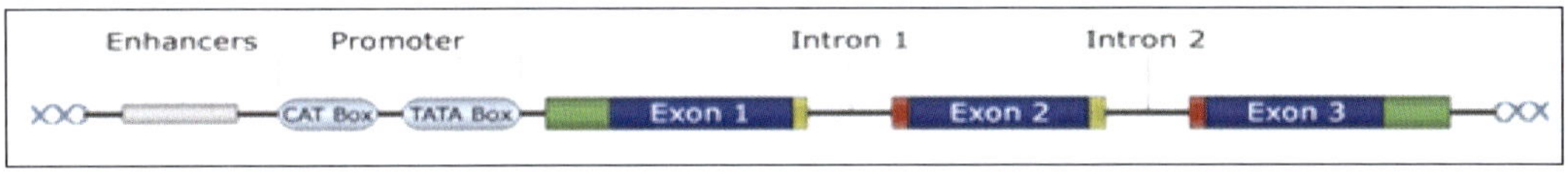

- Exons: Exons code for amino acids and collectively determine the amino acid sequence of the protein product. It is these portions of the gene that are represented in final mature mRNA molecule.
- Introns: Introns are portions of the gene that do not code for amino acids, and are removed (spliced) from the mRNA molecule before translation.

Gene Control Regions

- Start site: A start site for transcription.

- A promoter: A region a few hundred nucleotides 'upstream' of the gene (toward the 5' end). It is not transcribed into mRNA, but plays a role in controlling the transcription of the gene. Transcription factors bind to specific nucleotide sequences in the promoter region and assist in the binding of RNA polymerases.
- Enhancers: Some transcription factors (called activators) bind to regions called 'enhancers' that increase the rate of transcription. These sites may be thousands of nucleotides from the coding sequences or within an intron. Some enhancers are conditional and only work in the presence of other factors as well as transcription factors.
- Silencers: Some transcription factors (called repressors) bind to regions called 'silencers' that depress the rate of transcription.

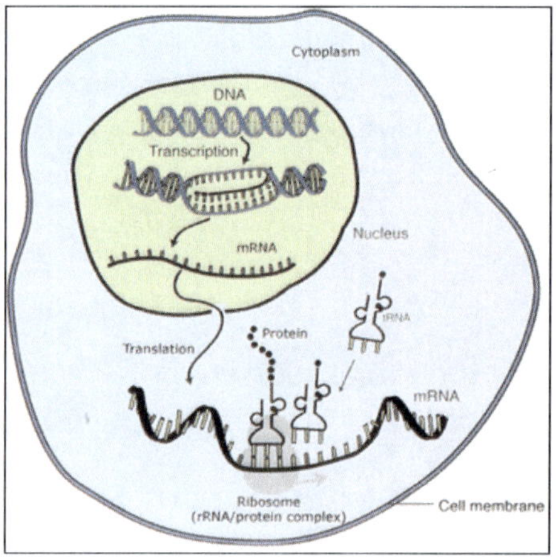

Transcription

Transcription is the first step of gene expression. During this process, the DNA sequence of a gene is copied into RNA.

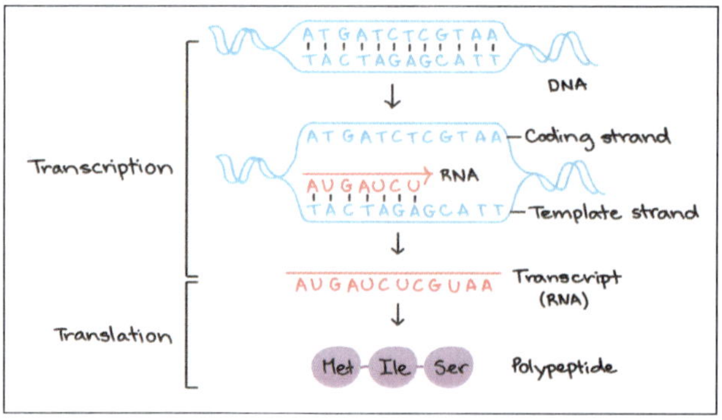

Before transcription can take place, the DNA double helix must unwind near the gene that is getting transcribed. The region of opened-up DNA is called a transcription bubble.

Transcription uses one of the two exposed DNA strands as a template; this strand is called the template strand. The RNA product is complementary to the template strand and is almost identical to the other DNA strand, called the nontemplate (or coding) strand. However, there is one important difference: in the newly made RNA, all of the T nucleotides are replaced with U nucleotides.

The site on the DNA from which the first RNA nucleotide is transcribed is called the +1 site, or the initiation site. Nucleotides that come before the initiation site are given negative numbers and said to be upstream. Nucleotides that come after the initiation site are marked with positive numbers and said to be downstream.

If the gene that's transcribed encodes a protein (which many genes do), the RNA molecule will be read to make a protein in a process called translation.

RNA Polymerase

RNA polymerases are enzymes that transcribe DNA into RNA. Using a DNA template, RNA polymerase builds a new RNA molecule through base pairing. For instance, if there is a G in the DNA template, RNA polymerase will add a C to the new, growing RNA strand.

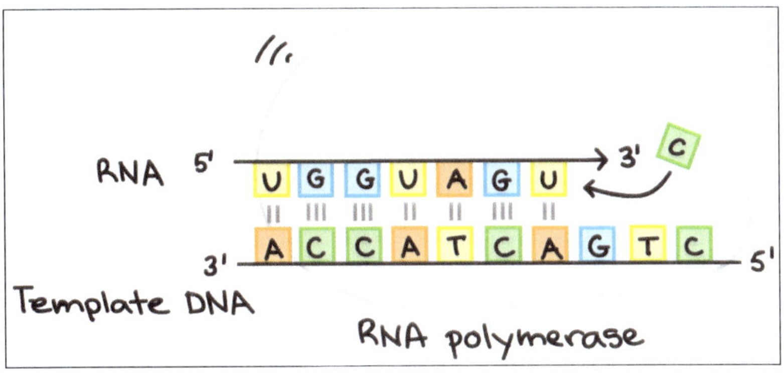

RNA polymerase always builds a new RNA strand in the 5' to 3' direction. That is, it can only add RNA nucleotides (A, U, C, or G) to the 3' end of the strand.

RNA polymerases are large enzymes with multiple subunits, even in simple organisms like bacteria. In addition, humans and other eukaryotes have three different kinds of RNA polymerases: I, II, and III. Each one specializes in transcribing certain classes of genes.

Transcription Initiation

To begin transcribing a gene, RNA polymerase binds to the DNA of the gene at a region called the promoter. Basically, the promoter tells the polymerase where to "sit down" on the DNA and begin transcribing.

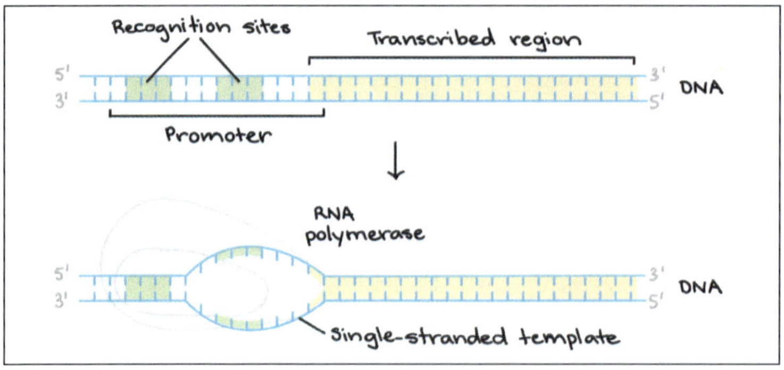

Each gene (or, in bacteria, each group of genes transcribed together) has its own promoter. A promoter contains DNA sequences that let RNA polymerase or its helper proteins attach to the DNA. Once the transcription bubble has formed, the polymerase can start transcribing.

Promoters in Bacteria

To get a better sense of how a promoter works, let's look an example from bacteria. A typical bacterial promoter contains two important DNA sequences, the -10 and -35 elements.

RNA polymerase recognizes and binds directly to these sequences. The sequences position the polymerase in the right spot to start transcribing a target gene, and they also make sure it's pointing in the right direction.

Once the RNA polymerase has bound, it can open up the DNA and get to work. DNA opening occurs at the -10 element, where the strands are easy to separate due to the many As and Ts (which bind to each other using just two hydrogen bonds, rather than the three hydrogen bonds of Gs and Cs).

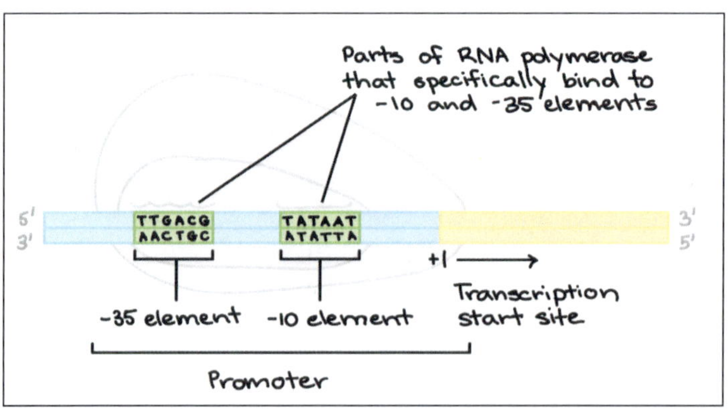

The -10 and the -35 elements get their names because they come 35 and 10 nucleotides before the initiation site (+1 in the DNA). The minus signs just mean that they are before, not after, the initiation site.

Promoters in Humans

In eukaryotes like humans, the main RNA polymerase in your cells does not attach directly to promoters like bacterial RNA polymerase. Instead, helper proteins called basal (general) transcription factors bind to the promoter first, helping the RNA polymerase in your cells get a foothold on the DNA.

Many eukaryotic promoters have a sequence called a TATA box. The TATA box plays a role much like that of the -10 element in bacteria. It's recognized by one of the general transcription factors, allowing other transcription factors and eventually RNA polymerase to bind. It also contains lots of As and Ts, which make it easy to pull the strands of DNA apart.

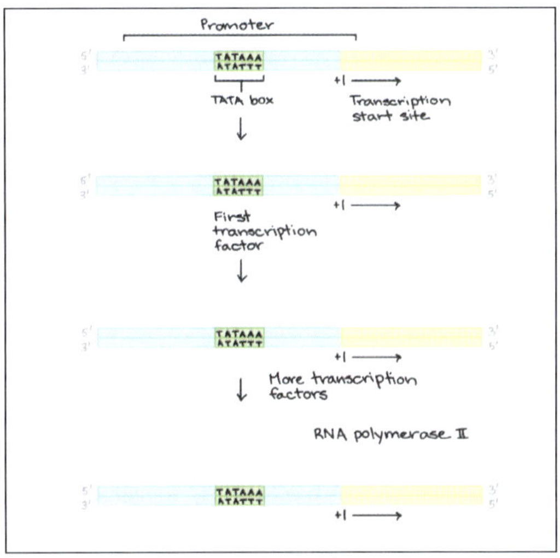

Elongation

Once RNA polymerase is in position at the promoter, the next step of transcription—elongation—can begin. Basically, elongation is the stage when the RNA strand gets longer, thanks to the addition of new nucleotides.

During elongation, RNA polymerase "walks" along one strand of DNA, known as the template strand, in the 3' to 5' direction. For each nucleotide in the template, RNA polymerase adds a matching (complementary) RNA nucleotide to the 3' end of the RNA strand.

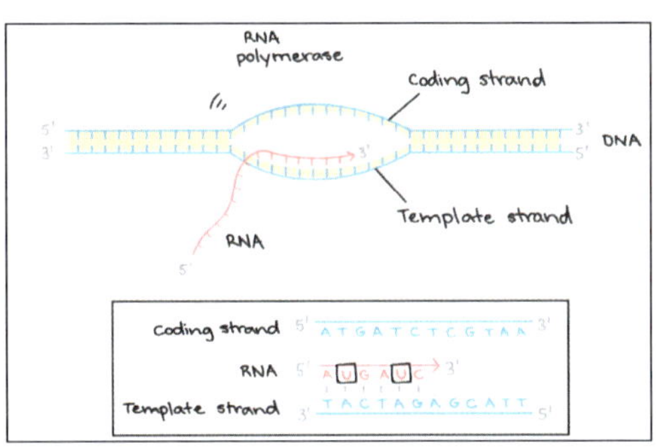

The RNA transcript is nearly identical to the non-template, or coding, strand of DNA. However, RNA strands have the base uracil (U) in place of thymine (T), as well as a slightly different sugar in the nucleotide. So, each T of the coding strand is replaced with a U in the RNA transcript.

The picture below shows DNA being transcribed by many RNA polymerases at the same time, each with an RNA "tail" trailing behind it. The polymerases near the start of the gene have short RNA tails, which get longer and longer as the polymerase transcribes more of the gene.

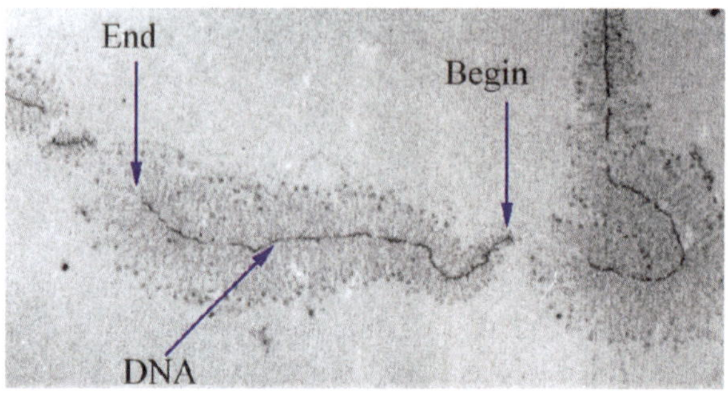

Transcription label.

Transcription Termination

RNA polymerase will keep transcribing until it gets signals to stop. The process of ending transcription is called termination, and it happens once the polymerase transcribes a sequence of DNA known as a terminator.

Termination in Bacteria

There are two major termination strategies found in bacteria: Rho-dependent and Rho-independent.

In Rho-dependent termination, the RNA contains a binding site for a protein called Rho factor. Rho factor binds to this sequence and starts "climbing" up the transcript towards RNA polymerase.

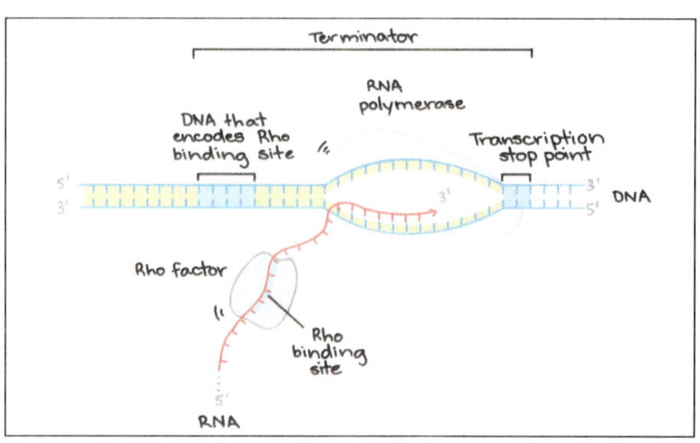

When it catches up with the polymerase at the transcription bubble, Rho pulls the RNA transcript and the template DNA strand apart, releasing the RNA molecule and ending transcription. Another sequence found later in the DNA, called the transcription stop point, causes RNA polymerase to pause and thus helps Rho catch up.

Rho-independent termination depends on specific sequences in the DNA template strand. As the RNA polymerase approaches the end of the gene being transcribed, it hits a region rich in C and G nucleotides. The RNA transcribed from this region folds back on itself, and the complementary C and G nucleotides bind together. The result is a stable hairpin that causes the polymerase to stall.

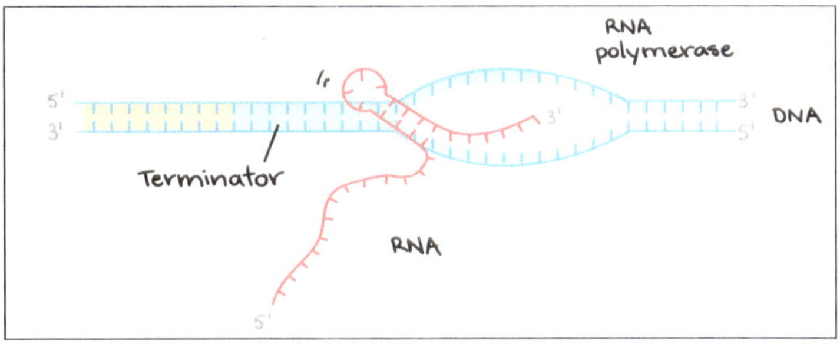

In a terminator, the hairpin is followed by a stretch of U nucleotides in the RNA, which match up with A nucleotides in the template DNA. The complementary U-A region of the RNA transcript forms only a weak interaction with the template DNA. This, coupled with the stalled polymerase, produces enough instability for the enzyme to fall off and liberate the new RNA transcript.

After termination, transcription is finished. An RNA transcript that is ready to be used in translation is called a messenger RNA (mRNA). In bacteria, RNA transcripts are ready to be translated right after transcription. In fact, they're actually ready a little sooner than that: translation may start while transcription is still going on.

Translation

Translation refers to the process of creating proteins from an mRNA template. The sequence of nucleotides on the RNA is translated into the amino acid sequence of proteins and this reaction is carried out by ribosomes. Ribosomes and tRNA dock on a mature mRNA transcript and recruit multiple enzymes in an energy intensive process that uses ATP as well as GTP.

When nucleic acids were discovered as the primary genetic material, one important question came to the fore. There are only 4 bases in nucleic acids whereas proteins are made of 20 amino acids. Therefore, it is not possible to have a direct one-to-one correlation between the sequences of nucleotides and amino acids. Even having two nucleotides code for a single amino acid is insufficient. Therefore, it was suggested that amino acids were coded by stretches of three nucleotides called codons. A series of experiments in the 1960s confirmed this hypothesis and also showed that these codons do not overlap with each other. In addition, since stretches of 3 nucleotides can give rise to a total of 64 codons, a single amino acid can be coded by multiple codons, a property that is called 'degeneracy'. Often, the difference between degenerate codons is the third base which is called the 'wobble position'. For instance, the amino acid serine can be coded by six codons of which four are: UCA, UCG, UCU or UCC. Similarly, phenylalanine can be represented by either UUU or UUC on the mRNA and leucine is coded by a total of six codons. This degeneracy is aided by the fact that the third nucleotide in every codon binds loosely to its corresponding tRNA, allowing unusual types of bases to pair with each other.

Out of the 64 codons made by various combinations of 4 nucleotides, 3 are stop codons, that signal the end of translation. These are UAA, UAG and UGA and are recognized by proteins called release factors rather than by tRNAs. When a ribosome encounters a stop codon, it dissociates from the mRNA through the enzymatic action of release factors.

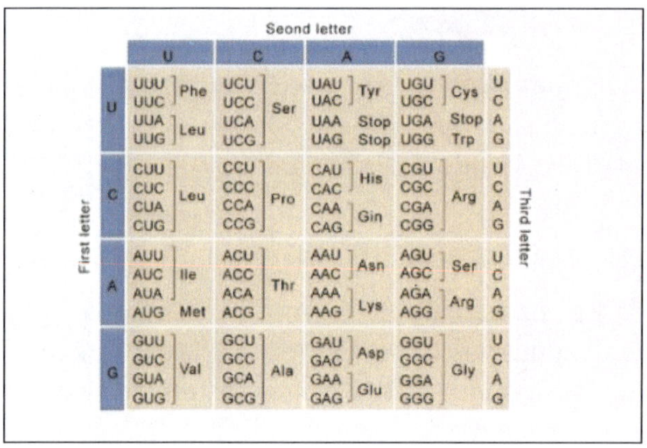

mRNA Untranslated Regions

The entire stretch of mature mRNA does not consist of codons that are translated into amino acid sequence of a protein. There is a 'cap' to the 5′ end of the RNA, two short stretches of untranslated regulatory regions abutting the coding sequence (the 5′ UTR and 3′ UTR), and a polyadenylate tail that can determine protein sequence without directly being translated.

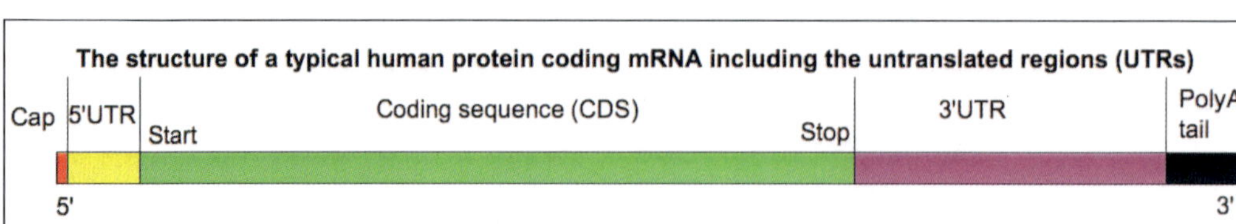

These regions are involved in the export of mRNA from the nucleus, protection from enzymatic degradation, and regulation of translational activity. They contain binding sites for proteins that can enhance or reduce translation, docking sites for ribosomes and other parts of the translation machinery, as well as enzymes that catalyze the degradation of mRNA when the protein requirement of the cell is met.

tRNA Structure and Function

Transfer RNA act as adapter molecules between mRNA and amino acids, bringing the appropriate amino acid to the ribosome based on mRNA codons. tRNAs contain a three-base anticodon that can recognize and bind with mRNA as well as act as a signal for the correct amino acid. The anticodon sequence is complementary to the mRNA codon, and runs in an antiparallel direction, allowing the two molecules to base pair with each other.

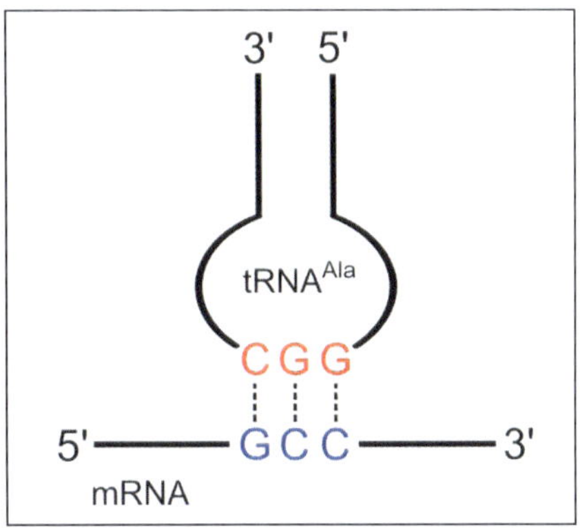

A group of enzymes called aminoacyl tRNA synthetases attach the appropriate amino acid to tRNA molecules based on their anticodon. There is one aminoacyl tRNA synthetase for each of the 20 amino acids and the enzyme can recognize all the anticodons that represent that particular amino acid. These enzymes use the energy from ATP to attach the amino acid to the last nucleotide on the 3′ end of tRNA. The tRNA is now considered to be 'charged' and can participate in the protein synthesizing reactions on the ribosome.

Ribosome Structure and Function

Ribosomes are macromolecular, multi-subunit structures containing RNA as well as protein and are the primary machines that drive protein synthesis. The structure of the ribosome derives primarily from it RNA component (ribosomal RNA or rRNA) and base-pairing with mRNA and tRNA is crucial to its function.

The ribosome contains two subunits and translation is initiated when the smaller subunit binds to sequences upstream of the coding sequence on the mRNA. Prokaryotic translation begins with the rRNA directly binding to the mRNA, whereas eukaryotic translation involves other proteins called initiation factors. The smaller subunit, along with some other proteins recruit the larger subunit of the ribosome, and translation begins.

Primarily, the ribosome contains three important regions – the P site, the A site and the E site – formed by the three-dimensional shape of the rRNA. The P site binds to the growing polypeptide, the A site anchors an incoming charged tRNA and after peptide bondformation, the tRNA binds briefly to the E site before leaving the ribosome.

Translation Mechanism

Translation proceeds in three stages – initiation, elongation and termination. Each of these is associated with different proteins and at every step, ATP and GTP are used as energy sources.

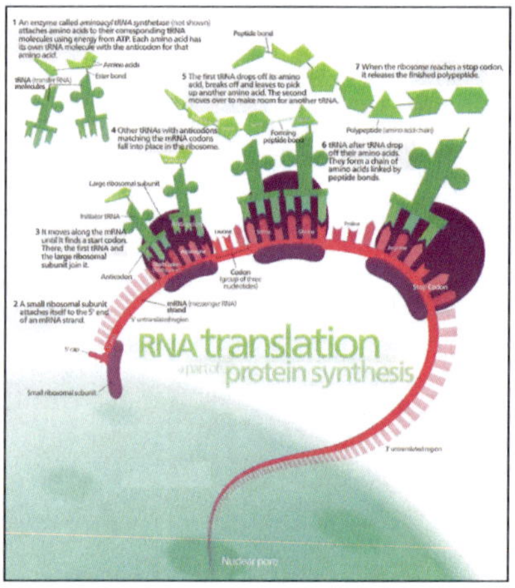

A single mRNA can be translated by multiple ribosomes in a process called translatome. These complexes were initially called ergosomes and are now called polysomes or polyribosomes.

Initiation

Translation begins with the mature mRNA transcript being exported from the nucleus and its 5′ cap being recognized by the smaller subunit of a ribosome. The ribosomal subunit, along with a special tRNA, scans the mRNA to find the start site for translation, which is often AUG – the codon for methionine. The sequences around the AUG start codon are also important and can determine how strongly an mRNA is translated. Initiation also involves the activity of a number of helper proteins called initiation factors, whose function is to ensure that the various parts of the translation machinery come together in an orderly manner.

The ribosome and initiator tRNA move slowly along the mRNA till the start codon is located. The structural features of this tRNA ensure that it is recognized by initiation factors and discriminated against by translation elongation factors. The methionine attached to this tRNA is also adapted to be exclusive for initiation – the amino group is modified to make N-formyl methionine, preventing it from participating in the elongation phase of translation.

Once the larger subunit of the ribosome arrives at the translation start site, and the tRNA is located at its P site, initiation is said to be complete.

Elongation

Binding of the initiator tRNA to the P site on the ribosome brings it into close contact with the amino or 'A' site of the ribosome, where the next codon awaits translation. New tRNAs carrying amino acids enter the ribosome at the A site. Through complementary base pairing and the energy from one molecule of GTP, the correct tRNA binds to the ribosome. Ribosomal RNA then catalyzes the formation of a peptide bond between the first and second amino acids where the first methionine appears to have been 'transferred' to the tRNA on the A site. Once the peptide bond is formed, the empty tRNA exits the ribosome and the ribosome itself moves ahead by exactly one codon, so that the tRNA on the A site moves into the P site. This also exposes the next codon to the A site, ready for the third tRNA. This process continues till a stop codon is reached.

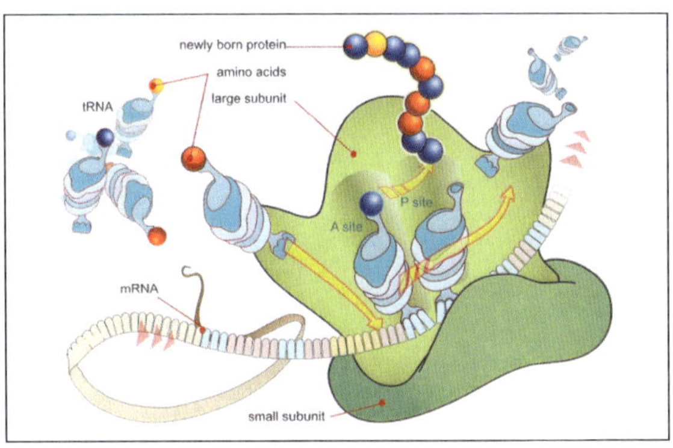

Termination

When a stop codon is present at the A site it is recognized by a set of proteins called release factors. They induce the ribosome to attach a molecule of water to the growing polypeptide chain, rather than another amino acid. This terminates the process of translation, and releases the polypeptide from the ribosome. The two subunits of the ribosome also dissociate from each other, ready for the next cycle of translation.

Translation on the Endoplasmic Reticulum

Translation can occur either on free ribosomes in the cytoplasm or on ribosomes present on the surface of the endoplasmic reticulum (ER). Translation begins in the cytoplasm for nearly all proteins. However proteins that are needed as internal membrane proteins or those that need to be secreted from the cell are targeted for further translation on the ER. These proteins contain a short stretch of hydrophobic residues called the signal peptide at the start of their sequence. As soon as these residues are translated, the signal peptide is recognized by specific proteins called signal-recognition particles, which can transport the entire ribosome and associated molecules to the membrane of the ER. The signal peptide gets embedded in the ER membrane and the rest of the protein is released into the interior space of the ER. The signal peptide is removed from proteins that need to be secreted from the cell, while those intended for internal membranes retain that short stretch providing a membrane anchor.

Occasionally, proteins needed inside organelles like mitochondria and chloroplasts are translated in the cytoplasm. These proteins are selectively transported into these organelles using specific proteins in an energy intensive process.

Antibiotic Targets

The differences between prokaryotic and eukaryotic translation machinery make them ideal drug targets for combating infections, while leaving the body's cells intact. These antibiotics include chloramphenicol, tetracycline, puromycin and erythromycin. However, since most animals also contain a rich inner biome of symbiotic bacteria, these antibiotics can also cause a number of side effects, including vitamin deficiencies.

Gene Mutation

A gene mutation is a permanent alteration in the DNA sequence that makes up a gene, such that the sequence differs from what is found in most people. Mutations range in size; they can affect anywhere from a single DNA building block (base pair) to a large segment of a chromosome that includes multiple genes.

Gene mutations can be classified in two major ways:

1. Hereditary Mutations: Hereditary mutations are inherited from a parent and are present throughout a person's life in virtually every cell in the body. These mutations are also called germline mutations because they are present in the parent's egg or sperm cells, which are also called germ cells.

When an egg and a sperm cell unite, the resulting fertilized egg cell receives DNA from both parents. If this DNA has a mutation, the child that grows from the fertilized egg will have the mutation in each of his or her cells.

2. Somatic Mutations: Acquired (or somatic) mutations occur at some time during a person's life and are present only in certain cells, not in every cell in the body. These changes can be caused by environmental factors such as ultraviolet radiation from the sun, or can occur if a mistake is made as DNA copies itself during cell division.

Acquired mutations in somatic cells (cells other than sperm and egg cells) cannot be passed on to the next generation.

De Novo Mutations

Genetic changes that are described as de novo (new) mutations can be either hereditary or somatic. In some cases, the mutation occurs in a person's egg or sperm cell but is not present in any of the person's other cells.

In other cases, the mutation occurs in the fertilized egg shortly after the egg and sperm cells unite. It is often impossible to tell exactly when a de novo mutation happened. As the fertilized egg divides, each resulting cell in the growing embryo will have the mutation.

De novo mutations may explain genetic disorders in which an affected child has a mutation in every cell in the body but the parents do not, and there is no family history of the disorder.

Mosaicism

Somatic mutations that happen in a single cell early in embryonic development can lead to a situation called mosaicism. These genetic changes are not present in a parent's egg or sperm cells, or in the fertilized egg, but happen a bit later when the embryo includes several cells.

As all the cells divide during growth and development, cells that arise from the cell with the altered gene will have the mutation, while other cells will not. Depending on the mutation and how many cells are affected, mosaicism may or may not cause health problems.

Disease-causing Gene Mutations

Most disease-causing gene mutations are uncommon in the general population. However, other genetic changes occur more frequently. Genetic alterations that occur in more than 1 percent of the population are called polymorphisms.

They are common enough to be considered a normal variation in the DNA. Polymorphisms are responsible for many of the normal differences between people such as eye color, hair color, and blood type. Although many polymorphisms have no negative effects on a person's health, some of these variations may influence the risk of developing certain disorders.

Types of Genetic Mutations

Because the genetic code contains the information to make the stuff of life, errors in an organism's DNA can have catastrophic consequences. Errors can happen during DNA replication if the wrong base pair is added to a DNA strand, if a base is skipped, or if an extra base is added.

Rarely, these errors may actually be helpful – the "mistaken" version of the DNA may work better than the original, or have an entirely new function! In that case, the new version may become more successful, and its carrier may outcompete carriers of the old version in the population. This spread of new traits throughout a population is how evolution progresses.

Silent Mutations and Redundant Coding

In some cases, genetic mutations may not have any effect at all on the end product of a protein. This is because, as seen in the table above, most amino acids are connected to more than one codon.

Glycine, for example, is coded for by the codons GGA, GGC, GGG, and GGU. A mutation resulting in the wrong nucleotide being used for the last letter of the glycine codon, then, would make no difference. A codon starting in "GG" would still code for glycine, no matter what letter was used last.

The use of multiple codons for the same amino acid is thought to be a mechanism evolved over time to minimize the chance of a small mutation causing problems for an organism.

Missense Mutation

In a missense mutation, the substitution of one base pair for an incorrect base pair during DNA replication results in the wrong amino acid being used in a protein.

This may have a small affect on an organism, or a large one – depending on how important the amino acid is to the function of its protein, and what protein is effected.

A missense mutation may result in an enzyme that almost as well as the normal version – or an enzyme that does not function at all.

Nonsense Mutation

A nonsense mutation occurs when the incorrect base pair is used during DNA replication – but where the resulting codon does not code for an incorrect amino acid.

Instead, this error creates a stop codon or another piece of information that is indecipherable to the cell. As a result, the ribosome stops working on that protein and all subsequent codons are not transcribed.

Nonsense mutations lead to incomplete proteins, which may function very poorly or not at all.

Deletion

In a deletion mutation, one or more DNA bases are not copied during DNA replication. Deletion mutations come in a huge range of sizes – a single base pair may be missing, or a large piece of a chromosome may be missing.

Smaller mutations are not always less harmful. The loss of just one or two bases can result in a frameshift mutation that impairs a crucial gene.

By contrast, larger deletion mutations may be fatal – or may only result in disability, as in DiGeorge Syndrome and other conditions that result from the deletion of part of a chromosome.

The reason for this is that DNA is very much like computer source code – one piece of code might be crucial for the system to turn on at all, while other pieces of code might just ensure that a website looks pretty or loads quickly.

Depending on the function of the piece of code that is deleted or otherwise mutated, a small change can have catastrophic consequences – or a seemingly large corruption of code one can result in a system that is just a bit glitchy.

Insertion

An insertion mutation occurs when one or more nucleotides is erroneously added to a growing DNA strand during DNA replication. On rare occasions, long stretches of DNA may be incorrectly added in the middle of a gene.

Like a missense mutation, the impact of this can vary. The addition of an unnecessary amino acid in a protein may make the protein only slightly less efficient; or it may cripple it.

Consider what would happen to your chair if you added a random piece of wood to it that the instructions did not call for. The results could vary a lot depending on the size, shape, and placement of the extra piece.

Duplication

A duplication mutation occurs when a segment of DNA is accidentally replicated two or more times. Like the other mutations listed above, these may have mild effects – or they may be catastrophic.

To imagine if your chair had two backs, two seats, or eight legs. A small duplication and the chair may still be useable, if a little odd-looking or uncomfortable. But if the chair had, for example, six seats attached to each other, it may rapidly become useless for its intended purpose.

Frameshift Mutation

A frameshift mutation is a subtype of insertion, deletion, and duplication mutations. In a frameshift mutation, one or two amino acids are deleted or inserted – resulting in a shifting of the "frame" which the ribosome uses to tell where one codon stops and the next begins.

This type of error can be especially dangerous because it causes all codons that occur after the error to be misread. Typically, every amino acid added to the protein after the frameshift mutation is wrong.

Imagine if you were reading a book – but at some point during the writing, a programming error happened such that every subsequent letter shifted one letter later in the alphabet.

A word that was supposed to read "letter" would suddenly become "mfuuft."

This is approximately what happens in a frameshift mutation.

Variation

Variation is any difference between cells, individual organisms, or groups of organisms of any species caused either by genetic differences (genotypic variation) or by the effect of environmental factors on the expression of the genetic potentials (phenotypic variation). Variation may be shown in physical appearance, metabolism, fertility, mode of reproduction, behaviour, learning and mental ability, and other obvious or measurable characters.

Genotypic variations are caused by differences in number or structure of chromosomes or by differences in the genes carried by the chromosomes. Eye colour, body form, and disease resistance are genotypic variations. Individuals with multiple sets of chromosomes are called polyploid; many common plants have two or more times the normal number of chromosomes, and new species may arise by this type of variation. A variation cannot be identified as genotypic by observation of the organism; breeding experiments must be performed under controlled environmental conditions to determine whether or not the alteration is inheritable.

Environmentally caused variations may result from one factor or the combined effects of several factors, such as climate, food supply, and actions of other organisms. Phenotypic variations also include stages in an organism's life cycle and seasonal variations in an individual. These variations do not involve any hereditary alteration and in general are not transmitted to future generations; consequently, they are not significant in the process of evolution.

Variations are classified either as continuous, or quantitative (smoothly grading between two extremes, with the majority of individuals at the centre, as height in human populations); or as discontinuous, or qualitative (composed of well-defined classes, as blood groups in man). A discontinuous variation with several classes, none of which is very small, is known as a polymorphic variation. The separation of most higher organisms into males and females and the occurrence of several forms of a butterfly of the same species, each coloured to blend with a different vegetation, are examples of polymorphic variation.

Genetic Variation

Genetic variation is a measure of the genetic differences that exist within a population. The genetic variation of an entire species is often called genetic diversity. Genetic variations are the differences in DNA segments or genes between individuals and each variation of a gene is called an allele. For example, a population with many different alleles at a single chromosome locus has a high amount of genetic variation. Genetic variation is essential for natural selection because natural selection can only increase or decrease frequency of alleles that already exist in the population.

Genetic variation is caused by:

- Mutation;
- Random mating between organisms;
- Random fertilization;
- Rrossing over (or recombination) between chromatids of homologous chromosomes during meiosis.

The last three of these factors reshuffle alleles within a population, giving offspring combinations which differ from their parents and from others.

Genetic variation in the shells of Donax variabilis.

In above figure, an enormous amount of phenotypic variation exists in the shells of Donax varabilis, otherwise known as the coquina mollusc. This phenotypic variation is due at least partly to genetic variation within the coquina population.

Evolution and Adaptation to the Environment

Variation allows some individuals within a population to adapt to the changing environment. Because natural selection acts directly only on phenotypes, more genetic variation within a population usually enables more phenotypic variation. Some new alleles increase an organism's ability to survive and reproduce, which then ensures the survival of the allele in the population. Other new alleles may be immediately detrimental (such as a malformed oxygen-carrying protein) and organisms carrying these new mutations will die out. Neutral alleles are neither selected for nor against and usually remain in the population. Genetic variation is advantageous because it enables some individuals and, therefore, a population, to survive despite a changing environment.

Low genetic diversity in the wild cheetah population.

In above figure, populations of wild cheetahs have very low genetic variation. Because wild cheetahs are threatened, their species has a very low genetic diversity. This low genetic diversity means they are often susceptible to disease and often pass on lethal recessive mutations; only about 5% of cheetahs survive to adulthood.

Geographic Variation

Some species display geographic variation as well as variation within a population. Geographic variation, or the distinctions in the genetic makeup of different populations, often occurs when populations are geographically separated by environmental barriers or when they are under selection pressures from a different environment. One example of geographic variation are clines: graded changes in a character down a geographic axis.

Sources of Genetic Variation

Gene duplication, mutation, or other processes can produce new genes and alleles and increase genetic variation. New genetic variation can be created within generations in a population, so a population with rapid reproduction rates will probably have high genetic variation. However, existing genes can be arranged in new ways from chromosomal crossing over and recombination in sexual reproduction. Overall, the main sources of genetic variation are the formation of new alleles, the altering of gene number or position, rapid reproduction, and sexual reproduction.

References

- Geneexpression-regulation: le.ac.uk, Retrieved 2 February, 2019
- Stages-of-transcription, transcription-of-dna-into-rna, gene-expression-central-dogma, biology, science: khanacademy.org, Retrieved April, 2019
- Translation: biologydictionary.net, Retrieved 13 June, 2019
- Gene-mutation: reliawire.com, Retrieved 17 March, 2019
- Genetic-code: biologydictionary.net, Retrieved 7 August, 2019
- Variation-biology, science: britannica.com, Retrieved January, 2019
- Genetic-Variation, Population-Genetics, The-Evolution-of-Populations, General-Biology: libretexts.org, Retrieved 22 July, 2019

Chapter 5

Gene Regulation

Gene regulation is a process that includes a wide range of mechanisms which helps cells to increase or decrease the production of specific gene products. It can be applied to different steps of gene expression such as transcription, translation and post translation. The chapter closely examines these key aspects of gene regulation to provide an extensive understanding of the subject.

Gene regulation is a label for the cellular processes that control the rate and manner of gene expression. A complex set of interactions between genes, RNA molecules, proteins (including transcription factors) and other components of the expression system determine when and where specific genes are activated and the amount of protein or RNA product produced.

Gene expression is primarily controlled at the level of transcription, largely as a result of binding of proteins to specific sites on DNA. In 1965 Francois Jacob, Jacques Monod, and Andre Lwoff shared the Nobel prize in medicine for their work supporting the idea that control of enzyme levels in cells is regulated by transcription of DNA. occurs through regulation of transcription, which can be either induced or repressed. These researchers proposed that production of the enzyme is controlled by an "operon," which consists a series of related genes on the chromosome consisting of an operator, a promoter, a regulator gene, and structural genes.

- The structural genes contain the code for the proteins products that are to be produced. Regulation of protein production is largely achieved by modulating access of RNA polymerase to the structural gene being transcribed.
- The promoter gene doesn't encode anything; it is simply a DNA sequence that is initial binding site for RNA polymerase.
- The operator gene is also non-coding; it is just a DNA sequence that is the binding site for the repressor.
- The regulator gene codes for synthesis of a repressor molecule that binds to the operator and blocks RNA polymerase from transcribing the structural genes.

The operator gene is the sequence of non-transcribable DNA that is the repressor binding site. There is also a regulator gene, which codes for the synthesis of a repressor molecule hat binds to the operator.

- Example of Inducible Transcription: The bacterium E. coli has three genes that encode for enzymes that enable it to split and metabolize lactose (a sugar in milk). The promoter is the site on DNA where RNA polymerase binds in order to initiate transcription. However, the enzymes are usually present in very low concentrations, because their transcription is inhibited by a repressor protein produced by a regulator gene. The repressor protein binds to the operator site and inhibits transcription. However, if lactose is present in the environment, it can bind to the repressor protein and inactivate it, effectively removing the blockade and enabling transcription of the messenger RNA needed for synthesis of these genes.

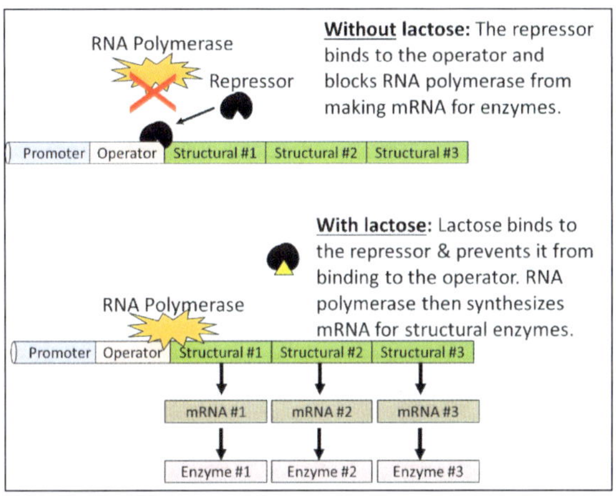

- Example of Repressible Transcription: E. coli need the amino acid tryptophan, and the DNA in E. coli also has genes for synthesizing it. These genes generally transcribe continuously since the bacterium needs tryptophan. However, if tryptophan concentrations are high, transcription is repressed (turned off) by binding to a repressor protein and activating it.

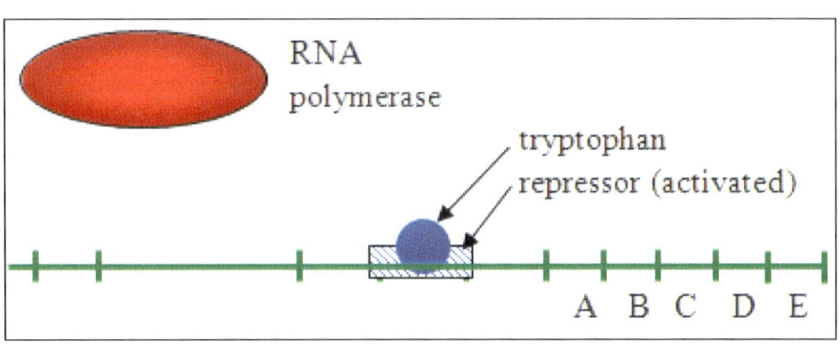

Control of Gene Expression in Eukaryotes

Eukaryotic cells have similar mechanisms for control of gene expression, but they are more complex. Consider, for example, that prokaryotic cells of a given species are all the same, but most eukaryotes are multicellular organisms with many cell types, so control of gene expression is much more complicated. Not surprisingly, gene expression in eukaryotic cells is controlled by a number of complex processes which are summarized by the following list:

- After fertilization, the cells in the developing embryo become increasingly specialized, largely by turning on some genes and turning off many others. Some cells in the pancreas, for example, are specialized to synthesize and secrete digestive enzymes, while other pancreatic cells (β-cells in the islets of Langerhans) are specialized to synthesis and secrete insulin. Each type of cell has a particular pattern of expressed genes. This differentiation into specialized cells occurs largely as a result of turning off the expression of most genes in the cell; mature cells may only use 3-5% of the genes present in the cell's nucleus.

- Gene expression in eukaryotes may also be regulated through by alterations in the packing of DNA, which modulates the access of the cell's transcription enzymes (e.g., RNA polymerase) to DNA. The illustration below shows that chromosomes have a complex structure. The DNA helix is wrapped around special proteins called histones, and this are wrapped into tight helical fibers. These fibers are then looped and folded into increasingly compact structures, which, when fully coiled and condensed, give the chromosomes their characteristic appearance in metaphase.

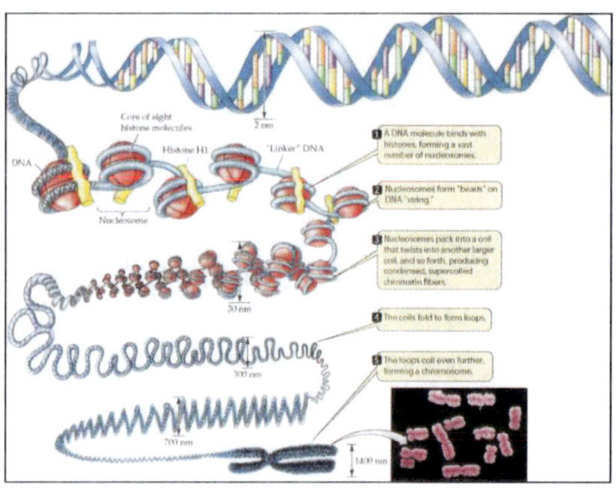

- Similar to the operons described above for prokaryotes, eukaryotes also use regulatory proteins to control transcription, but each eukaryotic gene has its own set of controls. In addition, there are many more *regulatory proteins* in eukaryotes and the interactions are much more complex.

- In eukaryotes transcription takes place within the membrane-bound nucleus, and the initial transcript is modified before it is transported from the nucleus to the cytoplasm for translation at the ribosome s. The initial transcript in eukaryotes has coding segments (exons) alternating with non-coding segments (introns). Before the mRNA leaves the nucleus, the introns are removed from the transcript by a process called RNA splicing, and extra nucleotides are added to the ends of the transcript; these non-coding "caps" and "tails" protect the mRNA from attack by cellular enzymes and aid in recognition by the ribosomes.

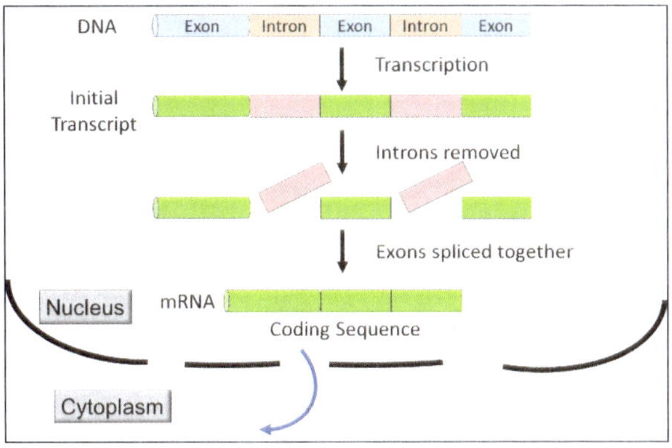

- Variation in the longevity of mRNA provides yet another opportunity for control of gene expression. Prokaryotic mRNA is very short-lived, but eukaryotic transcripts can last hours, or sometimes even weeks (e.g., mRNA for hemoglobin in the red blood cells of birds).
- The process of translation offers additional opportunities for regulation by many proteins. For example, the translation of hemoglobin mRNA is inhibited unless iron-containing heme is present in the cell.
- There are also opportunities for "post-translational" controls of gene expression in eukaryotes. Some translated polypeptides (proteins) are cut by enzymes into smaller, active final products. As illustrated in the figure below which depicts post-translational processing of the hormone insulin. Insulin is initially translated as a large, inactive precursor; a signal sequence is removed from the head of the precursor, and a large central portion (the C-chain) is cut away, leaving two smaller peptide chains which are then linked to each other by disulfide bridges. The smaller final form is the active form of insulin.

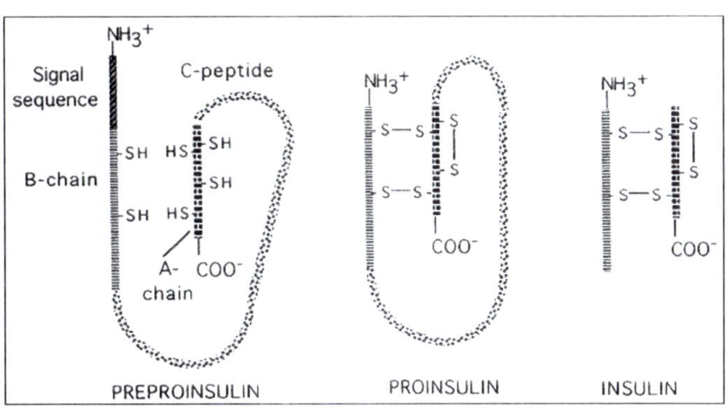

- Gene expression can also be modified by the breakdown of the proteins that are produced. For example, some of the enzymes involved in cell metabolism are broken down shortly after they are produced; this provides a mechanism for rapidly responding to changing metabolic demands.

- Gene expression can also be influenced by signals from other cells. There are many examples in which a signal molecule (e.g., a hormone) from one cell binds to a receptor protein on a target cell and initiates a sequence of biochemical changes (a signal transduction pathway) that result in changes within the target cell. These changes can include increased or decreased transcription as illustrated in the figure below.

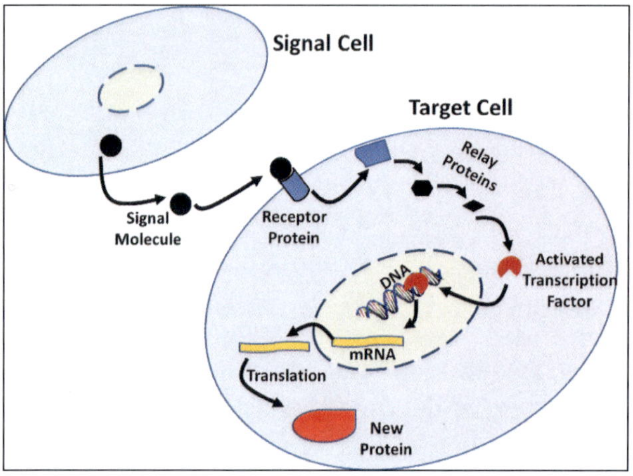

- The RNA Interference system (RNAi) is yet another mechanism by which cells control gene expression by shutting off translation of mRNA. RNAi can also be used to shut down translation of viral proteins when a cell is infected by a virus. The RNAi system also has the potential to be exploited therapeutically.

RNAi

Some RNA virus will invade cells and introduce double-stranded RNA which will use the cells machinery to make new copies of viral RNA and viral proteins. The cell's RNA interference system (RNAi) can prevent the viral RNA from replicating. First, an enzyme nicknamed "Dicer" chops any double-stranded RNA it finds into pieces that are about 22 nucleotides long. Next, protein complexes called RISC (RNA-induced Silencing Complex) bind to the fragments of double-stranded RNA, winds it, and then releases one of the strands, while retaining the other. The RISC-RNA complex will then bind to any other viral RNA with nucleotide sequences matching those on the RNA attached to the complex. This binding blocks translation of viral proteins at least partially, if not completely. The RNAi system could potentially be used to develop treatments for defective genes that cause disease. The treatment would involve making a double-stranded RNA from the diseased gene and introducing it into cells to silence the expression of that gene.

Transcriptional Regulation

Proteins called transcription factors, play a central role in regulating transcription. These important proteins help determine which genes are active in each cell of your body.

The enzyme RNA polymerase, which makes a new RNA molecule from a DNA template, must attach to the DNA of the gene. It attaches at a spot called the promoter.

In bacteria, RNA polymerase attaches right to the DNA of the promoter. You can see how this process works, and how it can be regulated by transcription factors, in the lac operon and trp operon videos.

In humans and other eukaryotes, there is an extra step. RNA polymerase can attach to the promoter only with the help of proteins called basal (general) transcription factors. They are part of the cell's core transcription toolkit, needed for the transcription of any gene.

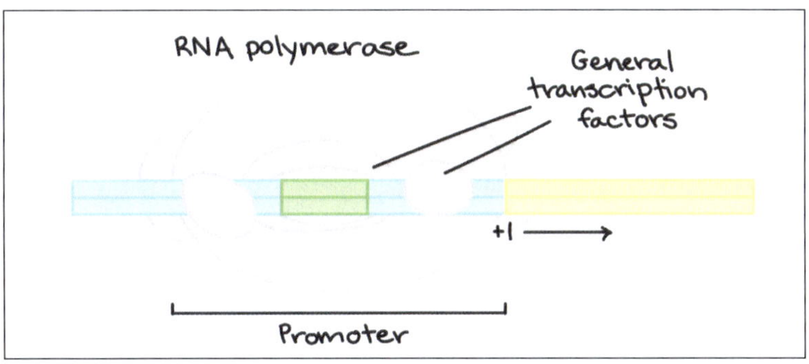

However, many transcription factors (including some of the coolest ones!) are not the general kind. Instead, there is a large class of transcription factors that control the expression of specific, individual genes. For instance, a transcription factor might activate only a set of genes needed in certain neurons.

Working of Transcription Factors

A typical transcription factor binds to DNA at a certain target sequence. Once it's bound, the transcription factor makes it either harder or easier for RNA polymerase to bind to the promoter of the gene.

Activators

Some transcription factors activate transcription. For instance, they may help the general transcription factors and/or RNA polymerase bind to the promoter, as shown in the diagram.

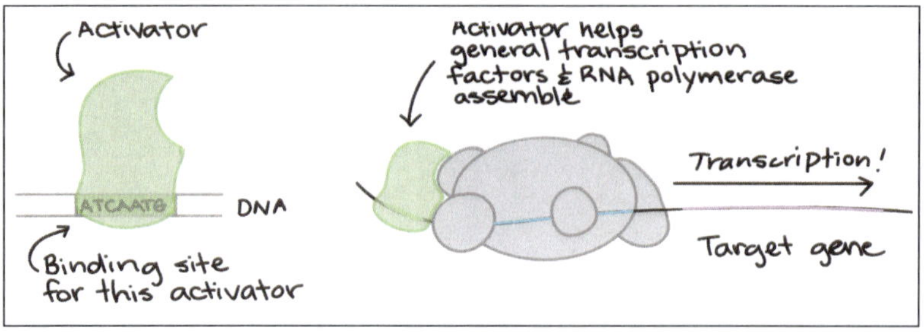

Repressors

Other transcription factors repress transcription. This repression can work in a variety of ways. As one example, a repressor may get in the way of the basal transcription factors or RNA polymerase, making it so they can't bind to the promoter or begin transcription.

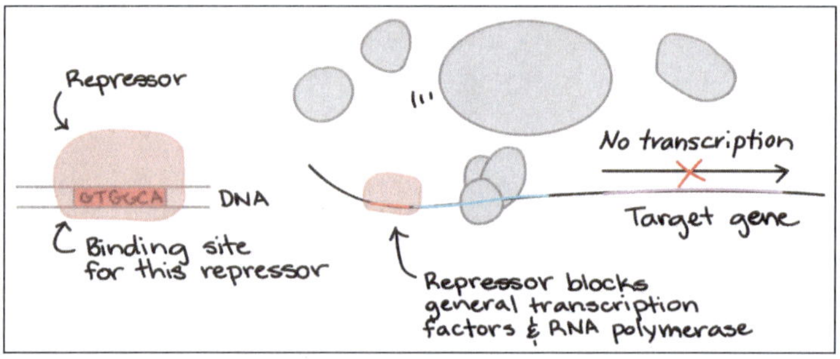

Binding Sites

The binding sites for transcription factors are often close to a gene's promoter. However, they can also be found in other parts of the DNA, sometimes very far away from the promoter, and still affect transcription of the gene.

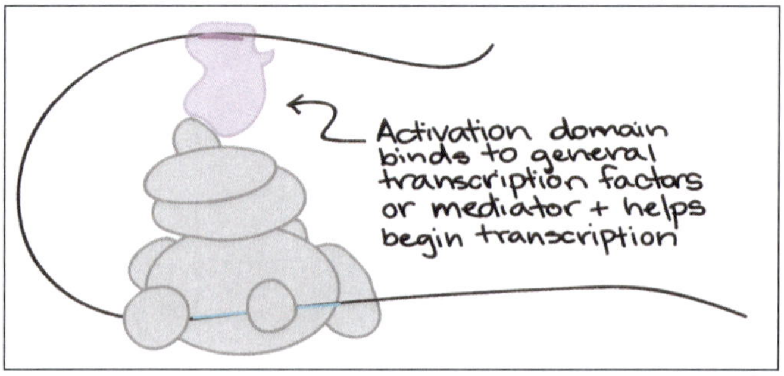

The flexibility of DNA is what allows transcription factors at distant binding sites to do their job. The DNA loops like cooked spaghetti to bring far-off binding sites and transcription factors close to general transcription factors or "mediator" proteins.

In the figure above, an activating transcription factor bound at a far-away site helps RNA polymerase bind to the promoter and start transcribing.

Turning Genes on in Specific Body Parts

Some genes need to be expressed in more than one body part or type of cell. For instance, suppose a gene needed to be turned on in your spine, skull, and fingertips, but not in the rest of your body. How can transcription factors make this pattern happen.

A gene with this type of pattern may have several enhancers (far-away clusters of binding sites for activators) or silencers (the same thing, but for repressors). Each enhancer or silencer may activate or repress the gene in a certain cell type or body part, binding transcription factors that are made in that part of the body.

Modular Mouse

As an example, let's consider a gene found in mice, called Tbx4. This gene is important for the development of many different parts of the mouse body, including the blood vessels and hind legs

During development, several well-defined enhancers drive Tbx4 expression in different parts of the mouse embryo. The diagram below shows some of the Tbx4 enhancers, each labeled with the body part where it produces expression.

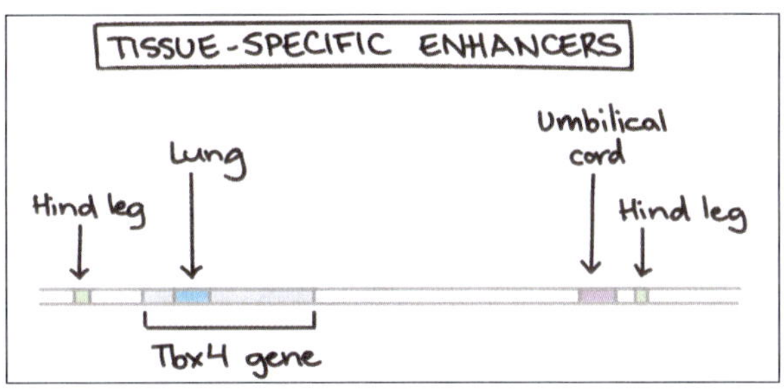

Evolution of Development

Enhancers like those of the Tbx4 gene are called tissue-specific enhancers: they control a gene's expression in a certain part of the body. Mutations of tissue-specific enhancers and silencers may play a key role in the evolution of body form.

Suppose that a mutation, or change in DNA, happened in the coding sequence of the Tbx4 gene. The mutation would inactivate the gene everywhere in the body, and a mouse without a normal copy would likely die. However, a mutation in an enhancer might just change the expression pattern a bit, leading to a new feature (e.g., a shorter leg) without killing the mouse.

Transcription Factors and Cellular "Logic"

Cells can detect information and combine it to determine the correct response—in much the same way that your calculator detects pushed buttons and outputs an answer.

We can see an example of this "molecular logic" when we consider how transcription factors regulate genes. Many genes are controlled by several different transcription factors, with a specific combination needed to turn the gene on; this is particularly true in eukaryotes and is sometimes called combinatorial regulation. For instance, a gene may be expressed only if activators A and B are present, and if repressor C is absent.

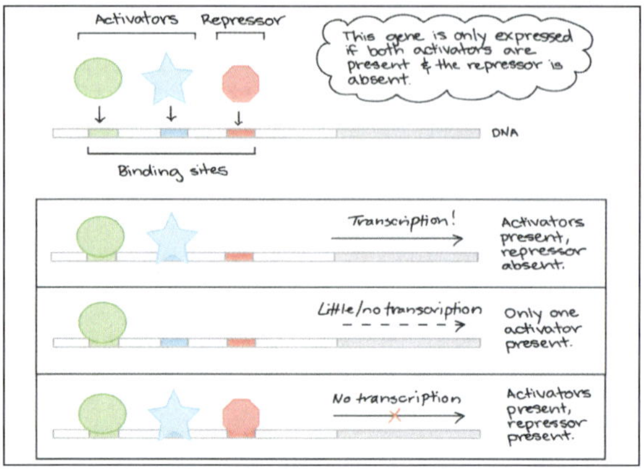

The use of multiple transcription factors to regulate a gene means that different sources of information can be integrated into a single outcome. For instance, imagine that:

- Activator A is present only in skin cells.
- Activator B is active only in cells receiving "divide now!" signals (growth factors) from neighbors.
- Repressor C is produced when a cell's DNA is damaged.

In this case, the gene would be "turned on" only in skin cells that are receiving division signals and have undamaged, healthy DNA. This pattern of regulation might make sense for a gene involved in cell division in skin cells. In fact, the loss of proteins similar to repressor C can lead to cancer.

Real-life combinatorial regulation can be a bit more complicated than this. For instance, many different transcription factors may be involved, or it may matter exactly how many molecules of a given transcription factor are bound to the DNA.

Post-transcriptional Regulation

Gene expression is often controlled primarily at the level of transcription. However, that doesn't mean transcription is the last chance for regulation. Later stages of gene expression can also be regulated, including:

- RNA processing, such as splicing, capping, and poly-A tail addition.

- Messenger RNA (mRNA) translation and lifetime in the cytosol.
- Protein modifications, such as addition of chemical groups.

Regulation of RNA Processing

When a eukaryotic gene is transcribed in the nucleus, the primary transcript (freshly made RNA molecule) isn't yet considered a messenger RNA. Instead, it's an "immature" molecule called a pre-mRNA.

The pre-mRNA has to go through some modifications to become a mature mRNA molecule that can leave the nucleus and be translated. These include splicing, capping, and addition of a poly-A tail, all of which can potentially be regulated – sped up, slowed down, or altered to result in a different product.

Alternative Splicing

Most pre-mRNA molecules have sections that are removed from the molecule, called introns, and sections that are linked or together to make the final mRNA, called exons. This process is called splicing.

In the process of alternative splicing, different portions of an mRNA can be selected for use as exons. This allows either of two (or more) mRNA molecules to be made from one pre-mRNA.

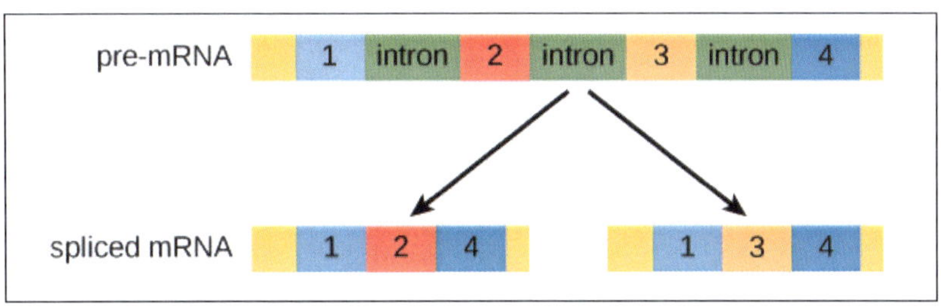

Eukaryotic Post-transcriptional Gene Regulation.

Alternative splicing is not a random process. Instead, it's typically controlled by regulatory proteins. The proteins bind to specific sites on the pre-mRNA and "tell" the splicing factors which exons should be used. Different cell types may express different regulatory proteins, so different exon combinations can be used in each cell type, leading to the production of different proteins.

Small Regulatory RNAs

Once an mRNA has left the nucleus, it may or may not be translated many times to make proteins. Two key determinants of how much protein is made from an mRNA are its "lifespan" (how long it floats around in the cytosol) and how readily the translation machinery, such as the ribosome, can attach to it.

A recently discovered class of regulators, called small regulatory RNAs, can control mRNA lifespan and translation.

MicroRNAs

MicroRNAs (miRNAs) were among the first small regulatory RNAs to be discovered. A miRNA is first transcribed as a long RNA molecule, which forms base pairs with itself and folds over to make a hairpin.

Next, the hairpin is chopped up by enzymes, releasing a small double-stranded fragment of about 2 nucleotides. One of the strands in this fragment is the mature miRNA, which binds to a specific protein to make an RNA-protein complex.

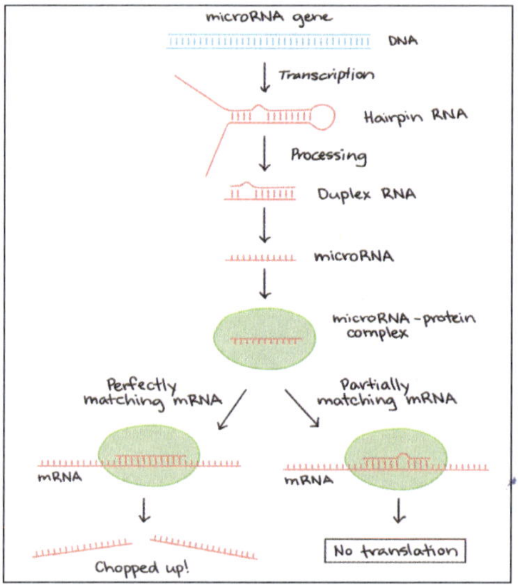

miRNA biogenesis.

The miRNA directs the protein complex to "matching" mRNA molecules (ones that form base pairs with the miRNA). When the RNA-protein complex binds:

- If the miRNA and its target match perfectly, an enzyme in the RNA-protein complex will typically chop the mRNA in half, leading to its breakdown.
- If the miRNA and its target have some mismatches, the RNA-protein complex may instead bind to the mRNA and keep it from being translated.

These are not the only ways that miRNAs inhibit expression of their targets, and scientists are still investigating their many modes of action·

What do miRNAs actually do in organisms? Their direct role is to reduce the expression of their target genes, but they may play this role to produce many different outcomes.

For instance, in mice, a specific miRNA plays a key role in the development and function of the vascular (circulatory) system. Mice without function of this miRNA had defects in heart development and were unable to survive. Changes in expression levels of miRNAs are also associated with human diseases, including various types of cancer and cardiac hypertrophy.

Translational Regulation

Gene expression is primarily regulated at the pre-transcriptional level, but there are a number of mechanisms for regulation of translation as well. One well-studied animal system is the iron-sensitive RNA-binding protein, which regulates the expression of genes involved in regulating intracellular levels of iron ions. Two of these genes, ferritin, which safely sequesters iron ions inside cells, and transferrin, which transports iron from the blood into the cell, both utilize this translational regulation system in a feedback loop to respond to intracellular iron concentration, but they react in opposite ways. The key interaction is between the iron response elements (IRE), which are sequences of mRNA that form short stem-loop structures, and IRE-BP, the protein that recognizes and binds to the IREs. In the case of the ferritin gene, the IRE sequences are situated upstream of the start codon. When there is high iron, the IRE-BP is inactive, and the stem-loop structures are melted and overrun by the ribosome, allowing translation of ferritin, which is an iron-binding protein. As the iron concentration drops, the IRE-BP is activated and binds around the IRE stem-loop structures, stabilizing them and preventing the ribosome from proceeding. This prevents the production of ferritin when there is little iron to bind.

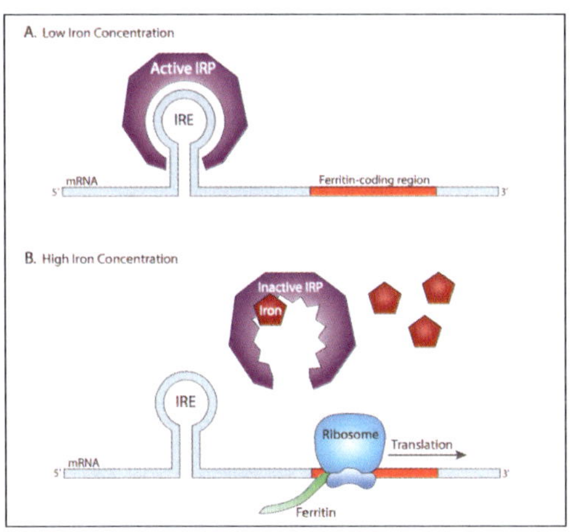

Pre-translational control of gene expression by iron-response protein (IRP), which binds to either the iron-response element (IRE), unless it has bound iron.

Transferrin also uses iron response elements and IRE-binding proteins, but in a very different mechanism. The IRE sequences of the transferrin gene are located downstream of the stop codon, and play no direct role in allowing or preventing translation.

However, when there is low intracellular iron and there is a need for more transferrin to bring iron into the cell, the IRE-BP is activated as in the previous case, and it binds to the IREs to stabilize the stem-loop structures. In this case; however, it prevents the 3' poly-A tail degradation that would normally occur over time. Once the poly-A tail is degraded, the rest of the mRNA is destroyed soon thereafter. the longer poly-A tails are associated with greater persistence in the cytoplasm, allowing more translation before they are destroyed. The IRE-BP system in this case externally prolongs the lifetime of the mRNA when that gene product is needed in higher amounts.

Since mRNA is a single-stranded nucleic acid and thus able to bind complementary sequence, it is not too surprising to find that one of the ways that a cell can regulate translation is using another piece of RNA. Micro RNAs (miRNAs) were discovered as very short (~20 nucleotides) non-protein-coding genes in the nematode, C. elegans. Since their initial discovery, hundreds have been found in various eukaryotes, including humans. The expression pattern of the miRNA genes is highly specific to tissue and developmental stage. Many are predicted to form stem-loop structures, and appear to hybridize to 3'-untranslated sequences of mRNA thus blocking initiation of translation on those mRNA molecules. They may also work through a mechanism similar to the siRNA, but there is clear evidence that mRNA levels are not necessarily altered by miRNA-directed translational control.

Another mechanism for translational control that uses small RNA molecules is RNA interference (RNAi). This was first discovered as an experimentally induced repression of translation when short double-stranded RNA molecules, a few hundred nucleotides in length and containing the same sequence as a target mRNA, were introduced into cells. The effect was dramatic: most of the mRNA with the target sequence was quickly destroyed. The current mechanistic model of RNAi repression is that first, the double-stranded molecules are cleaved by an endonuclease called Dicer, which cleaves with over-hanging single-stranded 3' ends. This allows the short fragments (siRNA, ~20nt long) to form a complex with several proteins (RISC, RNA-induced silencing complex). The RISC splits the double-stranded fragments into single strands, one of which is an exact complement to the mRNA. Because of the complementarity, this is a stable interaction, and the double-stranded region appears to signal an endonuclease to destroy the mRNA/siRNA hybrid.

The final method of controlling levels of gene expression is control after the fact, i.e., by targeted destruction of the gene product protein. While some proteins keep working until they fall apart, others are only meant for short-term use (e.g. to signal a short phase in the cell cycle) and need to be removed for the cell to function properly. Removal, in this sense, would be a euphemism for chopped up and recycled. The ubiquitin-proteasome system is a tag-and-destroy mechanism in which proteins that have outlived their usefulness are polyubiquitinated. Ubiquitin is a small (76 amino acids, ~5.6 kDa), highly conserved (96% between human and yeast sequences) eukaryotic protein that can be attached to other proteins through the action of three sequential enzymatic steps, each catalyzed by a different enzyme.

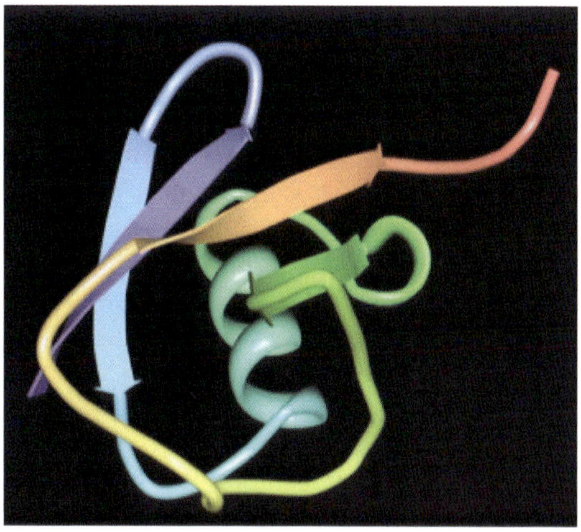

Ubiquitin. This 3D representation was generated from the file 1ubi (synthetic human ubiquitin) in the RCSB Protein Data Bank.

E1 activates the ubiquitin by combining it with ATP to make ubiquitin-adenylate, and then transfers the ubiquitin to itself via a cysteine thioester bond. Through a trans(thio) esterification reaction, the ubiquitin is then transferred to a cysteine in the E2 enzyme, also known as ubiquitin-conjugating enzyme. Finally, E3, or ubiquitin ligase, interacts with both E2-ubiquitin and the protein designated for destruction, transferring the ubiquitin to the target protein. After several rounds, the polyubiquitinated protein is send to the proteasome for destruction.

Mutations in E3 genes can cause a variety of human medical disorders such as the neurodevelopmental disorders Angelman syndrome, Hippel-Lindau syndrome, or the general growth disorder known as 3-M syndrome. Mechanisms linking malfunction in ubiquitination pathways and symptoms of these disorders are not currently known.

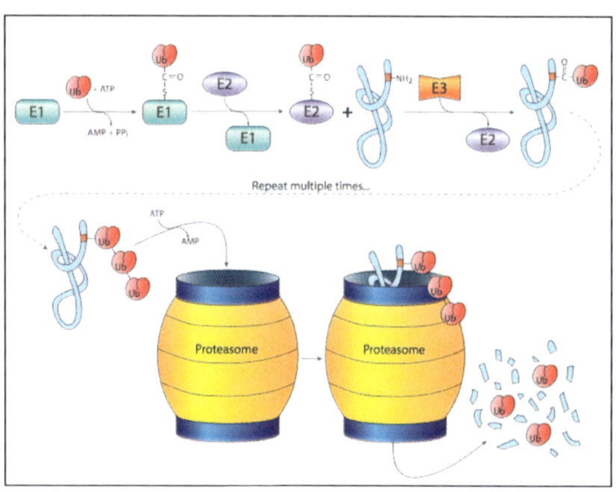

Polyubiquitination of a targeted protein (blue) requires three ubiquitinating enzymes, E1, E2, and E3. Once tagged, the protein is positioned in the proteasome by binding of the polyubiquitin tail to the outer surface of the proteasome. The proteasome then cleaves the protein into small polypeptides.

Proteasomes are very large protein complexes arranged as a four-layered barrel (the 20S subunit) capped by a regulatory subunit (19S) on each end. The two outer rings are each composed of 7 α subunits that function as entry gates to the central rings, each of which is composed of 7 β subunits, and which contain along the interior surface, 6 proteolytic sites. The 19S regulatory units control the opening and closing of the gates into the 20S catalytic barrel. The entire proteasome is sometimes referred to as a 26S particle.

A polyubiquitinated protein is first bound to the 19S regulatory unit in an ATP-dependent reaction (the 19S contains ATPase activity). 19S unit opens the gates of the 20S unit, possibly involving ATP hydrolysis, and guides the protein into the central proteolytic chamber. The protease activity of proteasomes is unique in that it is a threonine protease, and it cuts most proteins into regular 8-9 residue polypeptides, although this can vary.

Post-translational Regulation

RNA is transcribed, but must be processed into a mature form before translation can begin. This processing after an RNA molecule has been transcribed, but before it is translated into a protein, is called post-transcriptional modification. As with the epigenetic and transcriptional stages of processing, this post-transcriptional step can also be regulated to control gene expression in the cell. If the RNA is not processed, shuttled, or translated, then no protein will be synthesized.

RNA splicing, the first stage of post-transcriptional control.

In eukaryotic cells, the RNA transcript often contains regions, called introns, that are removed prior to translation. The regions of RNA that code for protein are called exons. After an RNA molecule has been transcribed, but prior to its departure from the nucleus to be translated, the RNA is processed and the introns are removed by splicing.

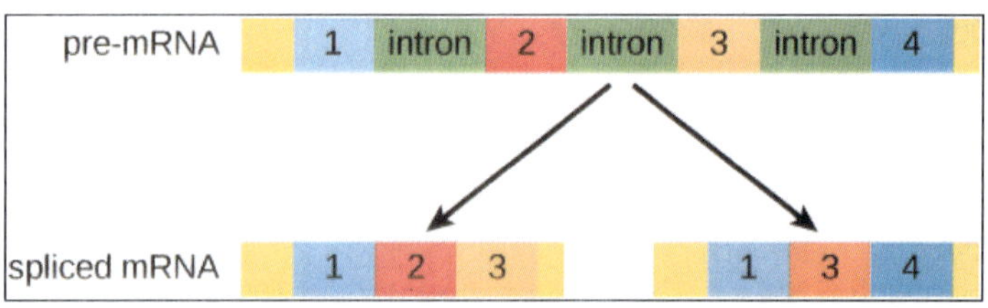

Pre-mRNA can be alternatively spliced to create different proteins.

Control of RNA Stability

Before the mRNA leaves the nucleus, it is given two protective "caps" that prevent the end of the strand from degrading during its journey. The 5′ cap, which is placed on the

5′ end of the mRNA, is usually composed of a methylated guanosine triphosphate molecule (GTP). The poly-A tail, which is attached to the 3′ end, is usually composed of a series of adenine nucleotides. Once the RNA is transported to the cytoplasm, the length of time that the RNA resides there can be controlled. Each RNA molecule has a defined lifespan and decays at a specific rate. This rate of decay can influence how much protein is in the cell. If the decay rate is increased, the RNA will not exist in the cytoplasm as long, shortening the time for translation to occur. Conversely, if the rate of decay is decreased, the RNA molecule will reside in the cytoplasm longer and more protein can be translated. This rate of decay is referred to as the RNA stability. If the RNA is stable, it will be detected for longer periods of time in the cytoplasm.

Binding of proteins to the RNA can influence its stability. Proteins, called RNA-binding proteins, or RBPs, can bind to the regions of the RNA just upstream or downstream of the protein-coding region. These regions in the RNA that are not translated into protein are called the untranslated regions, or UTRs. They are not introns (those have been removed in the nucleus). Rather, these are regions that regulate mRNA localization, stability, and protein translation. The region just before the protein-coding region is called the 5′ UTR, whereas the region after the coding region is called the 3′ UTR. The binding of RBPs to these regions can increase or decrease the stability of an RNA molecule, depending on the specific RBP that binds.

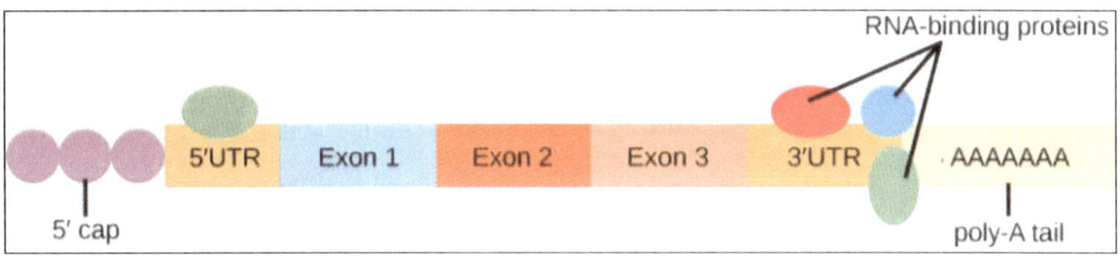

The protein-coding region of mRNA is flanked by 5′ and 3′ untranslated regions (UTRs). The presence of RNA-binding proteins at the 5′ or 3′ UTR influences the stability of the RNA molecule

RNA Stability and MicroRNAs

In addition to RBPs that bind to and control (increase or decrease) RNA stability, other elements called microRNAs can bind to the RNA molecule. These microRNAs, or miRNAs, are short RNA molecules that are only 21–24 nucleotides in length. The miRNAs are made in the nucleus as longer pre-miRNAs. These pre-miRNAs are chopped into mature miRNAs by a protein called dicer. Like transcription factors and RBPs, mature miRNAs recognize a specific sequence and bind to the RNA; however, miRNAs also associate with a ribonucleoprotein complex called the RNA-induced silencing complex (RISC). RISC binds along with the miRNA to degrade the target mRNA. Together, miRNAs and the RISC complex rapidly destroy the RNA molecule.

References

- Geneexpression-regulation: le.ac.uk, Retrieved 25 March, 2019
- DNA-Genetics: bumc.bu.edu, Retrieved 23 January, 2019
- Eukaryotic-transcription-factors, gene-regulation-in-eukaryotes, gene-regulation, biology, science: khanacademy.org, Retrieved 19 April, 2019
- Regulation-after-transcription, gene-regulation-in-eukaryotes, gene-regulation, biology, science: khanacademy.org, Retrieved 9 February, 2019
- Regulation-of-Translation, Molecules-and-Mechanisms, cells, Cell-and-Molecular-Biology: libretexts.org, Retrieved 11 June, 2019
- Post-translational-control-of-gene-expression: lumenlearning.com, Retrieved 2 May, 2019

Chapter 6

Genetic Engineering

The direct manipulation of an organism's genes using biotechnology is known as genetic engineering. Different organisms are modified using genetic engineering such as bacteria and virus. Genetic engineering is also involved in genetically modifying crops and providing gene therapy. All these diverse applications of gene therapy have been carefully analyzed in this chapter.

Genetic engineering is the direct altering of an organism's genome. This is achieved through manipulation of the DNA. Doing this is possible because DNA is like a universal language; all DNA for all organisms is made up of the same nucleotide building blocks. Thus, it is possible for genes from one organism to be read by another organism. In the cookbook analogy, this equates to taking a recipe from one organism's cookbook and putting into another cookbook. Now imagine that all cookbooks are written in the same language; thus, any recipe can be inserted and used in any other cookbook.

In practice, since DNA contains the genes to build certain proteins, by changing the DNA sequence, engineers are able to provide a new gene for a cell/organism to create a different protein. The new instructions may supplement the old instructions such that an extra trait is exhibited, or they may completely replace the old instructions such that a trait is changed.

Genetic Engineering Technique

The process for genetic engineering begins the same for any organism being modified.

1. Identify an organism that contains a desirable gene.
2. Extract the entire DNA from the organism.
3. Remove this gene from the rest of the DNA. One way to do this is by using a *restriction enzyme*. These enzymes search for specific nucleotide sequences where they will "cut" the DNA by breaking the bonds at this location.
4. Insert the new gene to an existing organism's DNA. This may be achieved through a number of different processes.

When modifying bacteria, the most common method for this final step is to add the isolated gene to a *plasmid*, a circular piece of DNA used by bacteria. This is done by "cutting" the plasmid with the same restriction enzyme that was used to remove the gene from the original DNA. The new gene can now be inserted into this opening in the plasmid and the DNA can be bonded back together using another enzyme called ligase. This process creates a *recombinant* plasmid. In this case, the recombinant plasmid is also referred to as a bacterial artificial chromosome (BAC).

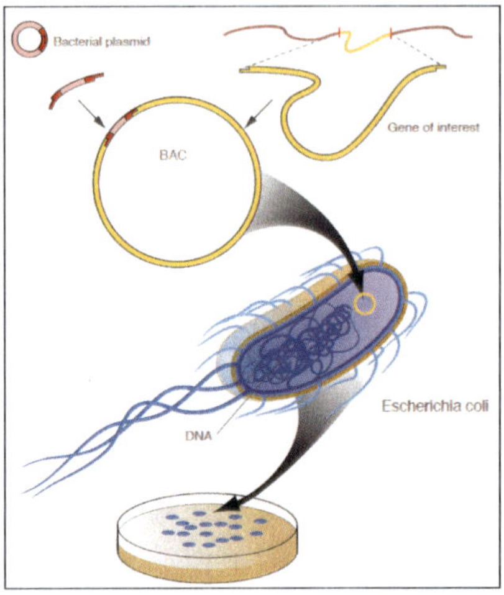

Building a recombinant plasmid to modify bacteria.

Once the recombinant DNA has been built, it can be passed to the organism to be modified. If modifying bacteria, this process is quite simple. The plasmid can be easily inserted into the bacteria where the bacteria treat it as their own DNA. For plant modification, certain bacteria such as *Agrobacterium tumefaciens* may be used because these bacteria permit their plasmids to be passed to the plant's DNA.

Genetically Modified Organism

A GMO, or genetically modified organism, is a plant, animal, microorganism or other organism whose genetic makeup has been modified in a laboratory using genetic engineering or transgenic technology. This creates combinations of plant, animal, bacterial and virus genes that do not occur in nature or through traditional crossbreeding methods.

Genetic modification affects many of the products we consume on a daily basis. As the number of GMOs available for commercial use grows every year, the Non-GMO Project works diligently to provide the most accurate, up-to-date standards for non-GMO verification.

In order for a product to be Non-GMO Project Verified, its inputs must be evaluated for compliance with our standard, which categorizes inputs into three risk levels.

Risk Level	Definition	Examples
High-Risk	The input is derived from, contains derivatives of, or is produced through a process involving organisms that are known to be genetically modified and commercially available.	Alfalfa, Canola, Corn, Cotton, Papaya, Soy, Sugar beet, Yellow summer squash / zucchini, Animal products, Microbes and enzymes, Potato
Low-Risk	The input is not derived from, does not contain derivatives of, or is not produced through a process involving organisms that are presently known to be genetically modified and commercially available.	Lentils, Spinach, Tomatoes, Sesame seeds, Avocados
Non-Risk	The input is not derived from biological organisms and not, therefore, susceptible to genetic modification.	
Monitored Risk	The Non-GMO Project carefully monitors the development of new genetically engineered products; we are currently tracking close to 100 products. Of those, we have included the following in our surveillance program, either because they will likely soon be widespread or because of known instances of contamination from GMOs.	Flax, Mustard, Rice, Wheat, Apple, Mushroom, Orange, Pineapple, Camelina (false flax), Salmon, Sugarcane, Tomato

Production

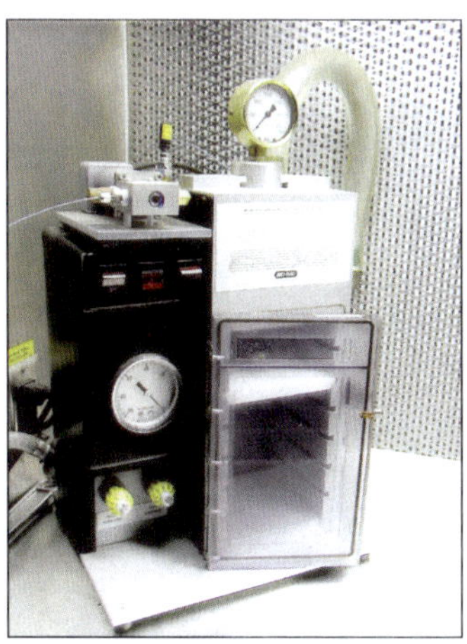

A gene gun uses biolistics to insert DNA into plant tissue.

Creating a genetically modified organism (GMO) is a multi-step process. Genetic engineers must isolate the gene they wish to insert into the host organism. This gene can be taken from a cell or artificially synthesized. If the chosen gene or the donor organism's genome has been well studied it may already be accessible from a genetic library. The gene is then combined with other genetic elements, including a promoter and terminator region and a selectable marker.

A number of techniques are available for inserting the isolated gene into the host genome. Bacteria can be induced to take up foreign DNA, usually by exposed heat shock or electroporation. DNA is generally inserted into animal cells using microinjection, where it can be injected through the cell's nuclear envelope directly into the nucleus, or through the use of viral vectors. In plants the DNA is often inserted using *Agrobacterium*-mediated recombination, biolistics or electroporation.

As only a single cell is transformed with genetic material, the organism must be regenerated from that single cell. In plants this is accomplished through tissue culture. In animals it is necessary to ensure that the inserted DNA is present in the embryonic stem cells. Further testing using PCR, Southern hybridization, and DNA sequencing is conducted to confirm that an organism contains the new gene.

Traditionally the new genetic material was inserted randomly within the host genome. Gene targeting techniques, which creates double-stranded breaks and takes advantage on the cells natural homologous recombination repair systems, have been developed to target insertion to exact locations. Genome editing uses artificially engineered nucleases that create breaks at specific points. There are four families of engineered nucleases: meganucleases, zinc finger nucleases, transcription activator-like effector nucleases (TALENs), and the Cas9-guideRNA system (adapted from CRISPR). TALEN and CRISPR are the two most commonly used and each has its own advantages. TALENs have greater target specificity, while CRISPR is easier to design and more efficient.

Fungi

Fungi can be used for many of the same processes as bacteria. For industrial applications, yeasts combines the bacterial advantages of being a single celled organism that is easy to manipulate and grow with the advanced protein modifications found in eukaryotes. They can be used to produce large complex molecules for use in food, pharmaceuticals, hormones and steroids. Yeast is important for wine production and as of 2016 two genetically modified yeasts involved in the fermentation of wine have been commercialized in the United States and Canada. One has increased malolactic fermentation efficiency, while the other prevents the production of dangerous ethyl carbamate compounds during fermentation. There have also been advances in the production of biofuel from genetically modified fungi.

Fungi, being the most common pathogens of insects, make attractive biopesticides. Unlike bacteria and viruses they have the advantage of infecting the insects by contact alone, although they are out competed in efficiency by chemical pesticides. Genetic engineering can improve virulence, usually by adding more virulent proteins, increasing infection rate or enhancing spore persistence. Many of the disease carrying vectors are susceptible to entomopathogenic fungi. An attractive target for biological control are mosquitos, vectors for a range of deadly diseases, including malaria, yellow fever and dengue fever. Mosquitos can evolve quickly so it becomes a balancing act of killing them before the *Plasmodium* they carry becomes the infectious disease, but not so fast that they become resistant to the fungi. By genetically engineering fungi like *Metarhizium anisopliae* and *Beauveria bassiana* to delay the development of mosquito infectiousness the selection pressure to evolve resistance is reduced. Another strategy is to add proteins to the fungi that block transmission of malaria or remove the *Plasmodium* altogether.

A mushroom has been gene edited to resist browning, giving it a longer shelf life. The process used CRISPR to knock out a gene that encodes polyphenol oxidase. As it didn't introduce any foreign DNA into the organism it was not deemed to be regulated under existing GMO frameworks and as such is the first CRISPR-edited organism to be approved for release. This has intensified debates as to whether gene-edited organisms should be considered genetically modified organisms and how they should be regulated.

Plants

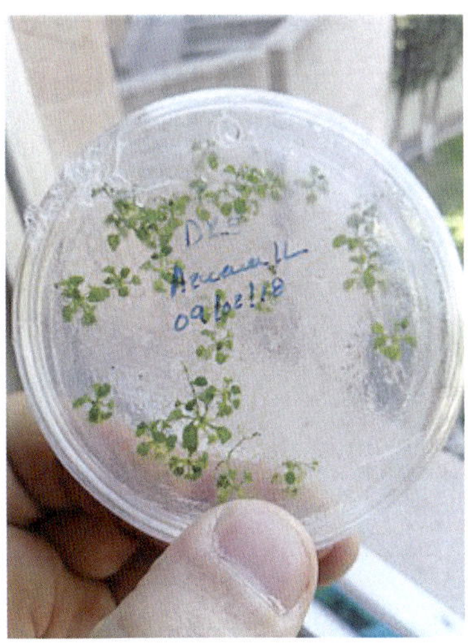

Tissue culture used to regenerate Arabidopsis thaliana.

Plants have been engineered for scientific research, to display new flower colors, deliver vaccines and to create enhanced crops. Many plants are pluripotent, meaning that a single cell from a mature plant can be harvested and under the right conditions can develop into a new plant. This ability can be taken advantage of by genetic engineers; by selecting for cells that have been successfully transformed in an adult plant a new plant can then be grown that contains the transgene in every cell through a process known as tissue culture.

Much of the advances in the field of genetic engineering has come from experimentation with tobacco. Major advances in tissue culture and plant cellular mechanisms for a wide range of plants has originated from systems developed in tobacco. It was the first plant to be altered using genetic engineering and is considered a model organism for not only genetic engineering, but a range of other fields. As such the transgenic tools and procedures are well established making tobacco one of the easiest plants to transform. Another major model organism relevant to genetic engineering is *Arabidopsis thaliana*. Its small genome and short life cycle makes it easy to manipulate and it contains many homologues to important crop species. It was the first plant sequenced, has a host of online resources available and can be transformed by simply dipping a flower in a transformed *Agrobacterium* solution.

In research, plants are engineered to help discover the functions of certain genes. The simplest way to do this is to remove the gene and see what phenotype develops compared to the wild type form. Any differences are possibly the result of the missing gene. Unlike mutagenisis, genetic engineering allows targeted removal without disrupting other genes in the organism. Some genes are only expressed in certain tissue, so reporter genes, like GUS, can be attached to the gene of interest allowing visualization of the location. Other ways to test a gene is to alter it slightly and then return it to the plant and see if it still has the same effect on phenotype. Other strategies include attaching the gene to a strong promoter and see what happens when it is over expressed, forcing a gene to be expressed in a different location or at different developmental stages.

Suntory "blue" rose.

Some genetically modified plants are purely ornamental. They are modified for flower color, fragrance, flower shape and plant architecture. The first genetically modified ornamentals commercialized altered color. Carnations were released in 1997, with the most popular genetically modified organism, a blue rose (actually lavender or mauve) created in 2004. The roses are sold in Japan, the United States, and Canada. Other genetically modified ornamentals include *Chrysanthemum* and *Petunia*. As well as increasing aesthetic value there are plans to develop ornamentals that use less water or are resistant to the cold, which would allow them to be grown outside their natural environments.

It has been proposed to genetically modify some plant species threatened by extinction to be resistant to invasive plants and diseases, such as the emerald ash borer in North American and the fungal disease, *Ceratocystis platani*, in European plane trees. The papaya ringspot virus devastated papaya trees in Hawaii in the twentieth century until transgenic papaya plants were given pathogen-derived resistance. However, genetic modification for conservation in plants remains mainly speculative. A unique concern is that a transgenic species may no longer bear enough resemblance to the original species to truly claim that the original species is being conserved. Instead, the transgenic species may be genetically different enough to be considered a new species, thus diminishing the conservation worth of genetic modification.

Crops

Wild type peanut (top) and transgenic peanut with Bacillus thuringiensis gene added (bottom) exposed to cornstalk borer larva.

Genetically modified crops are genetically modified plants that are used in agriculture. The first crops developed were used for animal or human food and provide resistance to certain pests, diseases, environmental conditions, spoilage or chemical treatments (e.g. resistance to a herbicide). The second generation of crops aimed to improve the quality, often by altering the nutrient profile. Third generation genetically modified crops could be used for non-food purposes, including the production of pharmaceutical agents, biofuels, and other industrially useful goods, as well as for bioremediation.

Kenyans examining insect-resistant transgenic Bacillus thuringiensis (Bt) corn.

There are three main aims to agricultural advancement; increased production, improved conditions for agricultural workers and sustainability. GM crops contribute by improving harvests through reducing insect pressure, increasing nutrient value and tolerating different abiotic stresses. Despite this potential, as of 2018, the commercialized crops are limited mostly to cash crops like cotton, soybean, maize and canola and the vast majority of the introduced traits provide either herbicide tolerance or insect resistance. Soybeans accounted for half of all genetically modified crops planted in 2014. Adoption by farmers has been rapid, between 1996 and 2013, the total surface area of land cultivated with GM crops increased by a factor of 100. Geographically though the spread has been uneven, with strong growth in the Americas and parts of Asia and little in Europe and Africa. Its socioeconomic spread has been more even, with approximately 54% of worldwide GM crops grown in developing countries in 2013. Although doubts have been raised, most studies have found growing GM crops to be beneficial to farmers through decreased pesticide use as well as increased crop yield and farm profit.

The majority of GM crops have been modified to be resistant to selected herbicides, usually a glyphosate or glufosinate based one. Genetically modified crops engineered to resist herbicides are now more available than conventionally bred resistant varieties; in the USA 93% of soybeans and most of the GM maize grown is glyphosate tolerant. Most currently available genes used to engineer insect resistance come from the *Bacillus thuringiensis* bacterium and code for delta endotoxins. A few use the genes that encode for vegetative insecticidal proteins. The only gene commercially used to provide insect protection that does not originate from *B. thuringiensis* is the Cowpea trypsin inhibitor (CpTI). CpTI was first approved for use cotton in 1999 and is currently undergoing trials in rice. Less than one percent of GM crops contained other traits, which include providing virus resistance, delaying senescence and altering the plants composition.

Golden rice is the most well known GM crop that is aimed at increasing nutrient value. It has been engineered with three genes that biosynthesise beta-carotene, a precursor of vitamin A, in the edible parts of rice. It is intended to produce a fortified food to be grown and consumed in areas with a shortage of dietary vitamin A, a deficiency which each year is estimated to kill 670,000 children under the age of 5 and cause an additional 500,000 cases of irreversible childhood blindness. The original golden rice produced 1.6μg/g of the carotenoids, with further development increasing this 23 times. In 2018 it gained its first approvals for use as food.

Golden rice compared to white rice.

Plants and plant cells have been genetically engineered for production of biopharmaceuticals in bioreactors, a process known as pharming. Work has been done with duckweed *Lemna minor*, the algae *Chlamydomonas reinhardtii* and the moss *Physcomitrella patens*. Biopharmaceuticals produced include cytokines, hormones, antibodies, enzymes and vaccines, most of which are accumulated in the plant seeds. Many drugs also contain natural plant ingredients and the pathways that lead to their production have been genetically altered or transferred to other plant species to produce greater volume. Other options for bioreactors are biopolymers and biofuels. Unlike bacteria, plants can modify the proteins post-translationally, allowing them to make more complex molecules.They also pose less risk of being contaminated. Therapeutics have been cultured in transgenic carrot and tobacco cells, including a drug treatment for Gaucher's disease.

Vaccine production and storage has great potential in transgenic plants. Vaccines are expensive to produce, transport and administer, so having a system that could produce them locally would allow greater access to poorer and developing areas. As well as purifying vaccines expressed in plants it is also possible to produce edible vaccines in plants. Edible vaccines stimulate the immune system when ingested to protect against certain diseases. Being stored in plants reduces the long-term cost as they can be disseminated without the need for cold storage, don't need to be purified and have long term stability. Also being housed within plant cells provides some protection from the gut acids upon digestion. However the cost of developing, regulating and containing transgenic plants is high, leading to most current plant-based vaccine development being applied to veterinary medicine, where the controls are not as strict.

Animals

The vast majority of genetically modified animals are at the research stage with the number close to entering the market remaining small. As of 2018 only three genetically modified animals have been approved, all in the USA. A goat and a chicken have been engineered to produce medicines and a salmon that has increased growth. Despite the differences and difficulties in modifying them, the end aims are much the same as for plants. GM animals are created for research purposes, production of industrial or therapeutic products, agricultural uses or improving their health. There is also a market for creating genetically modified pets.

Mammals

Some chimeras, like the blotched mouse shown, are created through genetic modification techniques like gene targeting.

The process of genetically engineering mammals is slow, tedious, and expensive. However, new technologies are making genetic modifications easier and more precise. The first transgenic mammals were produced by injecting viral DNA into embryos and then implanting the embryos in females. The embryo would develop and it would be hoped that some of the genetic material would be incorporated into the reproductive cells. Then researchers would have to wait until the animal reached breeding age and then offspring would be screened for presence of the gene in every cell. The development of the CRISPR-Cas9 gene editing system as a cheap and fast way of directly modifying germ cells, effectively halving the amount of time needed to develop genetically modified mammals.

Mammals are the best models for human disease, making genetic engineered ones vital to the discovery and development of cures and treatments for many serious diseases. Knocking out genes responsible for human genetic disorders allows researchers to study the mechanism of the disease and to test possible cures. Genetically modified mice have been the most common mammals used in biomedical research, as they are cheap and easy to manipulate. Pigs are also a good target as they have a similar body size and anatomical features, physiology, pathophysiological response and diet. Nonhuman primates are the most similar model organisms to humans, but there is less public acceptance towards using them as research animals. In 2009, scientists announced that they had successfully transferred a gene into a primate species (marmosets) for the first time. Their first research target for these marmosets was Parkinson's disease, but they were also considering amyotrophic lateral sclerosis and Huntington's disease.

Human proteins expressed in mammals are more likely to be similar to their natural counterparts than those expressed in plants or microorganisms. Stable expression has been accomplished in sheep, pigs, rats and other animals. In 2009 the first human biological drug produced from such an animal, a goat, was approved. The drug, ATryn, is an anticoagulant which reduces the probability of blood clots during surgery or childbirth and is extracted from the goat's milk. Human alpha-1-antitrypsin is another protein that has been produced from goats and is used in treating humans with this deficiency. Another medicinal area is in creating pigs with greater capacity for human organ transplants (xenotransplantation). Pigs have been genetically modified so that their organs can no longer carry retroviruses or have modifications to reduce the chance of rejection. Pig lungs from genetically modified pigs are being considered for transplantation into humans. There is even potential to create chimeric pigs that can carry human organs.

Livestock are modified with the intention of improving economically important traits such as growth-rate, quality of meat, milk composition, disease resistance and survival. Animals have been engineered to grow faster, be healthier and resist diseases. Modifications have also improved the wool production of sheep and udder health of cows. Goats have been genetically engineered to produce milk with strong spiderweb-like silk proteins in their milk. A GM pig called Enviropig was created with the capability of digesting plant phosphorus more efficiently than conventional pigs. They could reduce water pollution since they excrete 30 to 70% less phosphorus in manure. Dairy cows have been genetically engineered to produce milk that would be the same as human breast milk. This could potentially benefit mothers who cannot produce breast milk but want their children to have breast milk rather than formula. Researchers have also developed a genetically engineered cow that produces allergy-free milk.

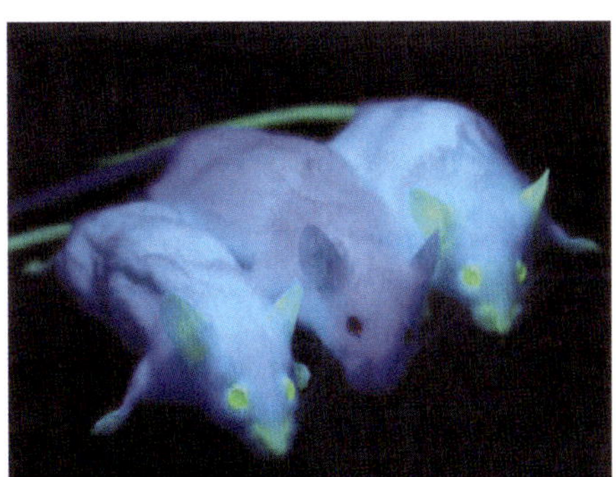

Mice expressing the green fluorescent protein.

Scientists have genetically engineered several organisms, including some mammals, to include green fluorescent protein (GFP), for research purposes. GFP and other similar reporting genes allow easy visualization and localization of the products of the genetic modification. Fluorescent pigs have been bred to study human organ transplants, regenerating ocular photoreceptor cells, and other topics. In 2011 green-fluorescent cats were created to help find therapies for HIV/AIDS and other diseases as feline immunodeficiency virus is related to HIV.

There have been suggestions that genetic engineering could be used to bring animals back from extinction. It involves changing the genome of a close living relative to resemble the extinct one and is currently being attempted with the passenger pigeon. Genes associated with the woolly mammoth have been added to the genome of an African Elephant, although the lead researcher says he has no intention of creating live elephants and transferring all the genes and reversing years of genetic evolution is a long way from being feasible. It is more likely that scientists could use this technology to conserve endangered animals by bringing back lost diversity or transferring evolved genetic advantages from adapted organisms to those that are struggling.

Humans

Gene therapy uses genetically modified viruses to deliver genes which can cure disease in humans. Although gene therapy is still relatively new, it has had some successes. It has been used to treat genetic disorders such as severe combined immunodeficiency, and Leber's congenital amaurosis. Treatments are also being developed for a range of other currently incurable diseases, such as cystic fibrosis, sickle cell anemia, Parkinson's disease, cancer, diabetes, heart disease and muscular dystrophy. These treatments only effect somatic cells, meaning any changes would not be inheritable. Germline gene therapy results in any change being inheritable, which has raised concerns within the scientific community.

In 2015, CRISPR was used to edit the DNA of non-viable human embryos. In November 2018, He Jiankui announced that he had edited the genomes of two human embryos, in an attempt to disable the *CCR5* gene, which codes for a receptor that HIV uses to enter cells. He said that twin girls, Lulu and Nana, had been born a few weeks earlier and that they carried functional copies of CCR5 along with disabled CCR5 (mosaicism) and were still vulnerable to HIV. The work was widely condemned as unethical, dangerous, and premature.

Fish

Genetically modified fish are used for scientific research, as pets and as a food source. Aquaculture is a growing industry, currently providing over half the consumed fish worldwide. Through genetic engineering it is possible to increase growth rates, reduce food intake, remove allergenic properties, increase cold tolerance and provide disease resistance. Fish can also be used to detect aquatic pollution or function as bioreactors.

Several groups have been developing zebrafish to detect pollution by attaching fluorescent proteins to genes activated by the presence of pollutants. The fish will then glow and can be used as environmental sensors. The GloFish is a brand of genetically modified fluorescent zebrafish with bright red, green, and orange fluorescent color. It was originally developed by one of the groups to detect pollution, but is now part of the ornamental fish trade, becoming the first genetically modified animal to become publicly available as a pet when in 2003 it was introduced for sale in the USA.

GM fish are widely used in basic research in genetics and development. Two species of fish, zebrafish and medaka, are most commonly modified because they have optically clear chorions (membranes in the egg), rapidly develop, and the one-cell embryo is easy to see and microinject with transgenic DNA. Zebrafish are model organisms for developmental processes, regeneration, genetics, behaviour, disease mechanisms and toxicity testing. Their transparency allows researchers to observe developmental stages, intestinal functions and tumour growth. The generation of transgenic protocols (whole organism, cell or tissue specific, tagged with reporter genes) has increased the level of information gained by studying these fish.

GM fish have been developed with promoters driving an over-production of growth hormone for use in the aquaculture industry to increase the speed of development and potentially reduce fishing pressure on wild stocks. This has resulted in dramatic growth enhancement in several species, including salmon, trout and tilapia. AquaBounty Technologies, a biotechnology company, have produced a salmon (called AquAdvantage salmon) that can mature in half the time as wild salmon. It obtained regulatory approval in 2015, the first non-plant GMO food to be commercialized. As of August 2017, GMO salmon is being sold in Canada. Sales in the US are expected to start in the second half of 2019.

Insects

In biological research, transgenic fruit flies (*Drosophila melanogaster*) are model organisms used to study the effects of genetic changes on development. Fruit flies are often preferred over other animals due to their short life cycle and low maintenance requirements. They also have a relatively simple genome compared to many vertebrates, with typically only one copy of each gene, making phenotypic analysis easy. *Drosophila* have been used to study genetics and inheritance, embryonic development, learning, behavior, and aging. The discovery of transposons, in particular the p-element, in *Drosophila* provided an early method to add transgenes to their genome, although this has been taken over by more modern gene-editing techniques.

Due to their significance to human health, scientists are looking at ways to control mosquitoes through genetic engineering. Malaria-resistant mosquitoes have been developed in the laboratory by inserting a gene that reduces the development of the malaria parasite and then use homing endonucleases to rapidly spread that gene throughout the male population (known as a gene drive). This approach has been taken further by using the gene drive to spread a lethal gene. In trials the populations of *Aedes aegypti* mosquitoes, the single most important carrier of dengue fever and Zika virus, were reduced by between 80% and by 90%. Another approach is to use a sterile insect technique, whereby males genetically engineered to be sterile out compete viable males, to reduce population numbers.

Other insect pests that make attractive targets are moths. Diamondback moths cause US$4 to $5 billion of damage each year worldwide. The approach is similar to the sterile technique tested on mosquitoes, where males are transformed with a gene that prevents any females born from reaching maturity. They underwent field trials in 2017. Genetically modified moths have previously been released in field trials. In this case a strain of pink bollworm that were sterilized with radiation were genetically engineered to express a red fluorescent protein making it easier for researchers to monitor them.

Silkworm, the larvae stage of *Bombyx mori,* is an economically important insect in sericulture. Scientists are developing strategies to enhance silk quality and quantity. There is also potential to use the silk producing machinery to make other valuable proteins. Proteins currently developed to be expressed by silkworms include; human serum albumin, human collagen α-chain, mouse monoclonal antibody and N-glycanase. Silkworms have been created that produce spider silk, a stronger but extremely difficult to harvest silk, and even novel silks.

Systems have been developed to create transgenic organisms in a wide variety of other animals. Chickens have been genetically modified for a variety of purposes. This includes studying embryo development, preventing the transmission of bird flu and providing evolutionary insights using reverse engineering to recreate dinosaur-like phenotypes.

Frog expressing Green fluorescent protein.

A GM chicken that produces the drug Kanuma, an enzyme that treats a rare condition, in its egg passed US regulatory approval in 2015. Genetically modified frogs, in particular *Xenopus laevis* and *Xenopus tropicalis*, are used in developmental biology research. GM frogs can also be used as pollution sensors, especially for endocrine disrupting chemicals. There are proposals to use genetic engineering to control cane toads in Australia.

The nematode *Caenorhabditis elegans* is one of the major model organisms for researching molecular biology. RNA interference (RNAi) was discovered in *C. elegans* and could be induced by simply feeding them bacteria modified to express double stranded RNA. It is also relatively easy to produce stable transgenic nematodes and this along with RNAi are the major tools used in studying their genes. The most common use of transgenic nematodes has been studying gene expression and localization by attaching reporter genes. Transgenes can also be combined with RNAi techniques to rescue phenotypes, study gene function, image cell development in real time or control expression for different tissues or developmental stages. Transgenic nematodes have been used to study viruses, toxicology, diseases, and to detect environmental pollutants.

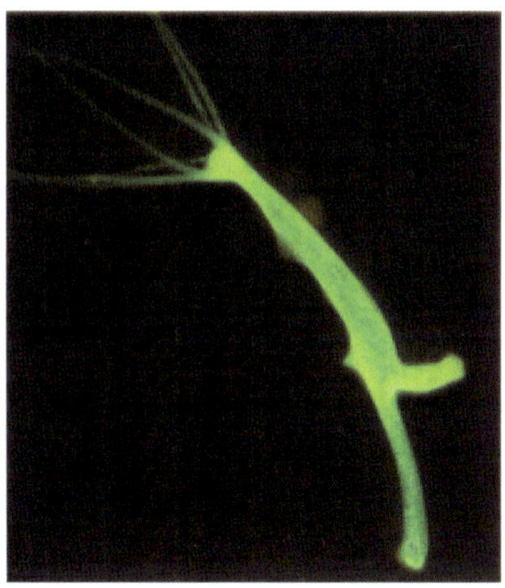

Transgenic Hydro expressing Green fluorescent protein.

The gene responsible for albinism in sea cucumbers has been found and used to engineer white sea cucumbers, a rare delicacy. The technology also opens the way to investigate the genes responsible for some of the cucumbers more unusual traits, including hibernating in summer, eviscerating their intestines, and dissolving their bodies upon death. Flatworms have the ability to regenerate themselves from a single cell. Until 2017 there was no effective way to transform them, which hampered research. By using microinjection and radiation scientists have now created the first genetically modified flatworms. The bristle worm, a marine annelid, has been modified. It is of interest due to its reproductive cycle being synchronized with lunar phases, regeneration capacity and slow evolution rate. Cnidaria such as *Hydra* and the sea anemone *Nematostella vectensis* are attractive model organisms to study the evolution of immunity and certain developmental processes. Other animals that have been genetically modified include snails, geckos, turtles, crayfish, oysters, shrimp, clams, abalone and sponges.

Genetically Modified Bacteria

Genetically modified bacteria were the first organisms to be modified in the laboratory, due to their simple genetics. These organisms are now used for several purposes, and are particularly important in producing large amounts of pure human proteins for use in medicine.

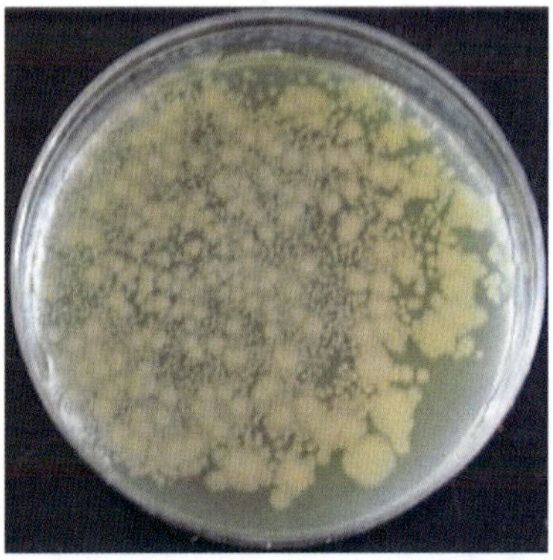

Bacteria transformed with pGLO under ambient light.

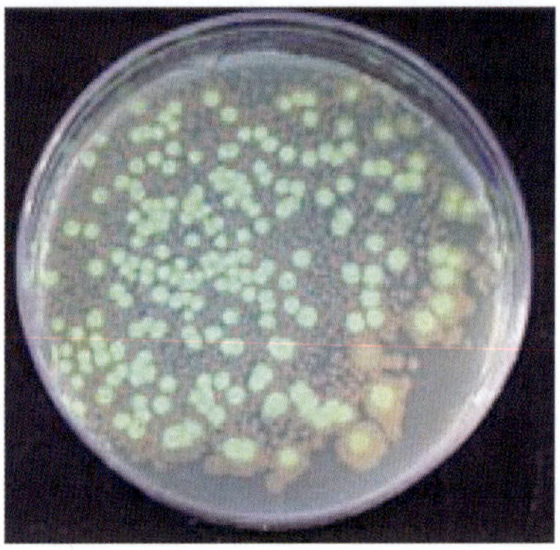

Bacteria transformed with pGLO visualised under ultraviolet light.

Bacteria were the first organisms to be genetically modified in the laboratory, due to the relative ease of modifying their chromosomes. This ease made them important tools for the creation of other GMOs. Genes and other genetic information from a wide range of organisms can be added to a plasmid and inserted into bacteria for storage and modification. Bacteria are cheap, easy to grow, clonal, multiply quickly, are relatively easy to transform, and can be stored at −80 °C almost indefinitely. Once a gene is isolated it can be stored inside the bacteria, providing an unlimited supply for research. The large number of custom plasmids make manipulating DNA excised from bacteria relatively easy.

Their ease of use has made them great tools for scientists looking to study gene function and evolution. Most DNA manipulation takes place within bacterial plasmids before being transferred to another host. Bacteria are the simplest model organism and most of our early understanding of molecular biology comes from studying *Escherichia coli*.

Scientist can easily manipulate and combine genes within the bacteria to create novel or disrupted proteins and observe the effect this has on various molecular systems. Researchers have combined the genes from bacteria and archaea, leading to insights on how these two diverged in the past. In the field of synthetic biology, they have been used to test various synthetic approaches, from synthesizing genomes to creating novel nucleotides.

Food

Bacteria have been used in the production of food for a long time, and specific strains have been developed and selected for that work on an industrial scale. They can be used to produce enzymes, amino acids, flavourings, and other compounds used in food production. With the advent of genetic engineering, new genetic changes can easily be introduced into these bacteria. Most food-producing bacteria are lactic acid bacteria, and this is where the majority of research into genetically engineering food-producing bacteria has gone. The bacteria can be modified to operate more efficiently, reduce toxic byproduct production, increase output, create improved compounds, and remove unnecessary pathways. Food products from genetically modified bacteria include alpha-amylase, which converts starch to simple sugars, chymosin, which clots milk protein for cheese making, and pectinesterase, which improves fruit juice clarity.

Industrial

Genetically modified bacteria are used to produce large amounts of proteins for industrial use. Generally the bacteria are grown to a large volume before the gene encoding the protein is activated. The bacteria are then harvested and the desired protein purified from them. The high cost of extraction and purification has meant that only high value products have been produced at an industrial scale.

Pharmaceutical Production

The majority of the industrial products from bacteria are human proteins for use in medicine. Many of these proteins are impossible or difficult to obtain via natural methods and they are less likely to be contaminated with pathogens, making them safer. Prior to recombinant protein products, several treatments were derived from cadavers or other donated body fluids and could transmit diseases. Indeed, transfusion of blood products had previously led to unintentional infection of haemophiliacs with HIV or hepatitis C; similarly, treatment with human growth hormone derived from cadaver pituitary glands may have led to outbreaks of Creutzfeldt–Jakob disease.

The first medicinal use of GM bacteria was to produce the protein insulin to treat diabetes. Other medicines produced include clotting factors to treat haemophilia, human growth hormone to treat various forms of dwarfism, interferon to treat some cancers, erythropoietin for anemic patients, and tissue plasminogen activator which dissolves blood clots. Outside of medicine they have been used to produce biofuels. There is interest in developing an extracellular expression system within the bacteria to reduce costs and make the production of more products economical.

Health

With greater understanding of the role that the micobiome plays in human health, there is the potential to treat diseases by genetically altering the bacteria to, themselves, be therapeutic agents. Ideas include altering gut bacteria so they destroy harmful bacteria, or using bacteria to replace or increase deficient enzymes or proteins. One research focus is to modify *Lactobacillus*, bacteria that naturally provide some protection against HIV, with genes that will further enhance this protection. The bacteria which generally cause tooth decay have been engineered to no longer produce tooth-corroding lactic acid. These transgenic bacteria, if allowed to colonize a person's mouth, could perhaps reduce the formation of cavities. Transgenic microbes have also been used in recent research to kill or hinder tumors, and to fight Crohn's disease.

If the bacteria do not form colonies inside the patient, the person must repeatedly ingest the modified bacteria in order to get the required doses. Enabling the bacteria to form a colony could provide a more long-term solution, but could also raise safety concerns as interactions between bacteria and the human body are less well understood than with traditional drugs. There are concerns that horizontal gene transfer to other bacteria could have unknown effects. As of 2018 there are clinical trials underway testing the efficacy and safety of these treatments.

Agriculture

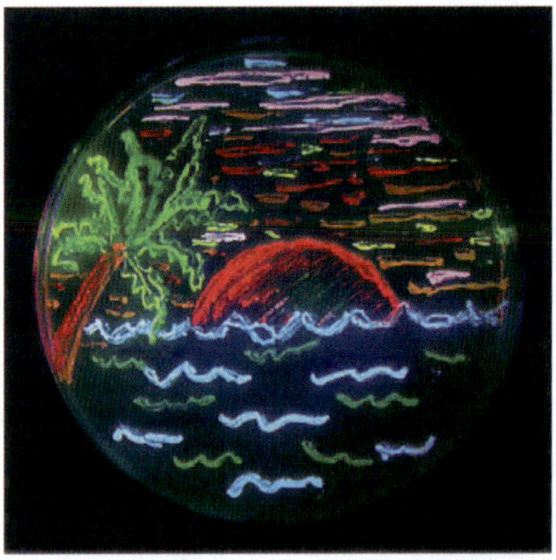

This artwork is made with bacteria modified to express 8 different colours of fluorescent proteins.

For over a century bacteria have been used in agriculture. Crops have been inoculated with Rhizobia (and more recently *Azospirillum*) to increase their production or to allow them to be grown outside their original habitat. Application of *Bacillus thuringiensis* (Bt) and other bacteria can help protect crops from insect infestation and plant diseases. With advances in genetic engineering, these bacteria have been manipulated for increased efficiency and expanded host range. Markers have also been added to aid in tracing the spread of the bacteria. The bacteria that naturally colonise certain crops have also been modified, in some cases to express the Bt genes responsible for pest resistance. *Pseudomonas* strains of bacteria cause frost damage by nucleating water into ice crystals around themselves. This led to the development of ice-minus bacteria, that have the ice-forming genes removed. When applied to crops they can compete with the ice-plus bacteria and confer some frost resistance.

Other uses

Other uses for genetically modified bacteria include bioremediation, where the bacteria are used to convert pollutants into a less toxic form. Genetic engineering can increase the levels of the enzymes used to degrade a toxin or to make the bacteria more stable under environmental conditions. GM bacteria have also been developed to leach copper from ore, clean up mercury pollution and detect arsenic in drinking water. Bioart has also been created using genetically modified bacteria. In the 1980s artist Jon Davis and geneticist Dana Boyd converted the Germanic symbol for femininity (ψ) into binary code and then into a DNA sequence, which was then expressed in *Escherichia coli*. This was taken a step further in 2012, when a whole book was encoded onto DNA. Paintings have also been produced using bacteria transformed with fluorescent proteins.

Genetically Modified Virus

A genetically modified virus is a virus that has been altered or generated using biotechnology methods, and remains capable of infection. Genetic modification involves the directed insertion, deletion, artificial synthesis or change of nucleotide bases in viral genomes. Genetically modified viruses are mostly generated by the insertion of foreign genes intro viral genomes for the purposes of biomedical, agricultural, bio-control, or technological objectives. The terms genetically modified virus and genetically engineered virus are used synonymously.

General usage

Genetically modified viruses are generated through genetic modification, which involves the directed insertion, deletion, artificial synthesis, or change of nucleotide sequences in viral genomes using biotechnological methods. While most dsDNA viruses have single monopartite genomes, many RNA viruses have multipartitegenomes, it is not necessary for all parts of a viral genome to be genetically modified for the virus to be considered a genetically modified virus. Infectious viruses capable of infection that are generated through artificial gene synthesis of all, or part of their genomes (for example based on inferred historical sequences) may also be considered as genetically modified viruses. Viruses that are changed solely through the action of spontaneous mutations, recombination or reassortment events (even in experimental settings), are not generally considered to be genetically modified viruses.

Viruses are generally modified so they can be used as vectors for inserting new genetic information into a host organism or altering its preexisting genetic material. This can be achieved in at least three processes:

1. Integration of all, or parts, of a viral genome into the host's genome (e.g. into its chromosomes). When the whole genetically modified viral genome is integrated it is then referred to as a genetically modified provirus. Where DNA or RNA which that has been packaged as part of a virus particle, but may not necessarily contain any viral genes, becomes integrated into a hosts genome this process is known as transduction.

2. Maintenance of the viral genome within host cells but not as an integrated part of the host's genome.

3. Where genes necessary for genome editing have been placed into the viral genome using biotechnology methods, editing of the host's genome is possible. This process does not require the integration of viral genomes into the host's genome.

None of these three processes are mutually exclusive. Where only process 2. occurs and it results in the expression of a genetically modified gene this will often be referred to as a transient expression approach.

The capacity to infect host cells or tissues is a necessary requirement for all applied uses of genetically modified viruses. However, a capacity for viral transmission (the transfer of infections *between* host individuals), is either not required or is considered undesirable for most applications. Only in a small minority of proposed uses is viral transmission considered necessary or desirable, an example is transmissible vaccines. This is because transmissibility considerably complicates to efforts monitor, control, or contain the spread of viruses.

Health Applications

Gene Therapy

Gene therapy uses genetically modified viruses to deliver genes that can cure diseases in human cells.These viruses can deliver DNA or RNA genetic material to the targeted cells. Gene therapy is also used by inactivating mutated genes that are causing the disease using viruses.

Viruses that have been used for gene therapy are, adenovirus, lentivirus, retrovirus and the herpes simplex virus.The most common virus used for gene delivery come from adenoviruses as they can carry up to 7.5kb of foreign DNA and infect a relatively broad range of host cells, although they have been know to elicit immune responses in the host and only provide short term expression. Other common vectors are adeno-associated viruses, which have lower toxicity and longer term expression, but can only carry about 4kb of DNA. Herpes simplex viruses is a promising vector, have a carrying capacity of over 30kb and provide long term expression, although it is less efficient at gene delivery than other vectors. The best vectors for long term integration of the gene into the host genome are retroviruses, but their propensity for random integration is problematic. Lentiviruses are a part of the same family as retroviruses with the advantage of infecting both dividing and non-dividing cells, whereas retroviruses only target dividing cells. Other viruses that have been used as vectors include alphaviruses, flaviviruses, measles viruses, rhabdoviruses, Newcastle disease virus, poxviruses, and picornaviruses.

Although primarily still at trial stages, it has had some successes. It has been used to treat inherited genetic disorders such as severe combined immunodeficiency rising from adenosine deaminase deficiency (ADA-SCID), although the development of leukemia in some ADA-SCID patients along with the death of Jesse Gelsinger in another trial set back the development of this approach for many years. In 2009 another breakthrough was achieved when an eight year old boy with Leber's congenital amaurosis regained normal eyesight and in 2016 GlaxoSmithKline gained approval to commercialise a gene therapy treatment for ADA-SCID. As of 2018, there are a substantial number of clinical trials underway, including treatments for hemophilia, glioblastoma, chronic granulomatous disease, cystic fibrosis and various cancers.Although some successes, gene therapy is still considered a risky technique and studies are still undergoing to ensure safety and effectiveness.

Cancer Treatment

Another potential use of genetically modified viruses is to alter them so they can directly treat diseases. This can be through expression of protective proteins or by directly targeting infected cells. In 2004, researchers reported that a genetically modified virus that exploits the selfish behaviour of cancer cells might offer an alternative way of killing tumours. Since then, several researchers have developed genetically modified oncolytic viruses that show promise as treatments for various types of cancer.

Vaccines

Most vaccines consist of viruses that have been attenuated, disabled, weakened or killed in some way so that their virulent properties are no longer effective. Genetic engineering could theoretically be used to create viruses with the virulent genes removed. In 2001, it was reported that genetically modified viruses can possibly be used to develop vaccines against diseases such as, AIDS, herpes, dengue fever and viral hepatitis by using a proven safe vaccine virus, such as adenovirus, and modify its genome to have genes that code for immunogenic proteins that can spike the immune systems response to then be able to fight the virus. Genetic engineered viruses should not have reduced infectivity, invoke a natural immune response and there is no chance that they will regain their virulence function, which can occur with some other vaccines. As such they are generally considered safer and more efficient than conventional vaccines, although concerns remain over non-target infection, potential side effects and horizontal gene transfer to other viruses. Another approach is to use vectors to create novel vaccines for diseases that have no vaccines available or the vaccines that are do not work effectively, such as AIDS, malaria, and tuberculosis. Vector-based vaccines have already been approved and many more are being developed.

Heart Pacemaker

In 2012, US researchers reported that they injected a genetically modified virus into the heart of pigs. This virus inserted into the heart muscles a gene called Tbx18 which enabled heartbeats. The researchers forecast that one day this technique could be used to restore the heartbeat in humans who would otherwise need electronic pacemakers.

Genetically Modified Viruses intended for use in the Environment

Animals

In Spain and Portugal, by 2005 rabbits had declined by as much as 95% over 50 years due diseases such as myxomatosis, rabbit haemorrhagic disease and other causes. This in turn caused declines in predators like the Iberian lynx, a critically endangered species. In 2000 Spanish researchers investigated a genetically modified virus which might have protected rabbits in the wild against myxomatosis and rabbit haemorrhagic disease. However, there was concern that such a virus might make its way into wild populations in areas such as Australia and create a population boom. Rabbits in Australia are considered to be such a pest that land owners are legally obliged to control them.

Genetically modified viruses that make the target animals infertile through immunocontraception have been created as well as others that target the developmental stage of the animal. There are concerns over virus containment and cross species infection.

Trees

Since 2009 genetically modified viruses expressing spinach defensin proteins have been field trialed in Florida (USA). The virus infection of orange trees aims to combat citrus greening disease, that had reduced orange production in Florida 70% since 2005. A permit application has been pending since February 13th 2017 (USDA 17-044-101r) to extend the experimental use permit to an area of 513,500 acres, this would make it the largest permit of this kind ever issued by the USDA Biotechnology Regulatory Services.

Insect Allies Program

In 2016 DARPA, an agency of the USA department of defense, announced a tender for contracts to develop genetically modified plant viruses for an approach involving their dispersion into the environment using insects. The work plan stated that:-

"Plant viruses hold significant promise as carriers of gene editing circuitry and are a natural partner for an insect-transmitted delivery platform."

The motivation provided for the program is to ensure food stability by protecting agricultural food supply and commodity crops:

> "By leveraging the natural ability of insect vectors to deliver viruses with high host plant specificity, and combining this capability with advances in gene editing, rapid enhancement of mature plants in the field can be achieved over large areas and without the need for industrial infrastructure."

Despite its name, the "Insect Allies" program is to a large extent a viral program, developing viruses that would essentially perform gene editing of crops in already-planted fields. The genetically modified viruses described in the work plan and other public documents are of a class of genetically modified viruses subsequently termed HEGAAs (horizontal environmental gene alteration agents). The Insect Allies program is scheduled to run from 2017 to 2021 with contracts being executed by three consortia. There are no plans to release the genetically modified viruses into the environment, with testing of the full insect dispersed system occurring in greenhouses.

Concerns have been expressed about how this program and any data it generates will impact biological weapon control and agricultural coexistence, though there has also been support for its stated motives.

Technological Applications

Lithium-ion Batteries

In 2009, MIT scientists created a genetically modified virus has been used to construct a more environmentally friendly lithium-ion battery. The battery was constructed by genetically engineering different viruses such as, the E4 bacteriophage and the M13 bacteriophage, to be used as a cathode. This was done by editing the genes of the virus that code for the protein coat. The protein coat is edited to coat itself in iron phosphate to be able to adhere to highly conductive carbon-nanotubes. The viruses that have been modified to have a multifunctional protein coat can be used as a nano-structured cathode with causes ionic interactions with cations. Allowing the virus to be used as a small battery. Angela Blecher, the scientist who led the MIT research team on the project, says that the battery is powerful enough to be used as a rechargeable battery, power hybrid electric cars, and a number of personal electronics. While both the E4 and M13 viruses can infect and replicate within their bacterial host, it unclear if they retain this capacity after being part of a battery.

Safety Concerns and Regulation

Bio-hazard Research Limitations

The National Institute of Health declared a research funding moratorium on select Gain-of-Function virus research in January 2015. Questions about a potential escape of a modified virus from a biosafety lab and the utility of dual-use-technology, dual use research of concern (DURC), prompted the NIH funding policy revision.

GMO Lentivirus Incident

A scientist claims she was infected by a genetically modified virus while working for Pfizer. In her federal lawsuit she says she has been intermittently paralyzed by the Pfizer-designed virus. "McClain, of Deep River, suspects she was inadvertently exposed, through work by a former Pfizer colleague in 2002 or 2003, to an engineered form of the lentivirus, a virus similar to the one that can lead to acquired immune deficiency syndrome, or AIDS." The court found that McClain failed to demonstrate that her illness was caused by exposure to the lentivirus, but also that Pfizer violated whistleblower laws.

Genetically Modified Food

Genetically modified food (or GM food) is food produced from plants or animals whose DNA has been altered through genetic engineering. These genetically modified organisms are often called GMOs for short.

Process

Genetically engineered organisms are generated and tested in the laboratory for desired qualities. The most common modification is to add one or more genes to an organism's genome. Less commonly, genes are removed or their expression is increased or silenced or the number of copies of a gene is increased or decreased.

Once satisfactory strains are produced, the producer applies for regulatory approval to field-test them, called a "field release". Field-testing involves cultivating the plants on farm fields or growing animals in a controlled environment. If these field tests are successful, the producer applies for regulatory approval to grow and market the crop. Once approved, specimens (seeds, cuttings, breeding pairs, etc.) are cultivated and sold to farmers. The farmers cultivate and market the new strain. In some cases, the approval covers marketing but not cultivation.

According to the USDA, the number of field releases for genetically engineered organisms has grown from four in 1985 to an average of about 800 per year. Cumulatively, more than 17,000 releases had been approved through September 2013.

Crops

Fruits and Vegetables

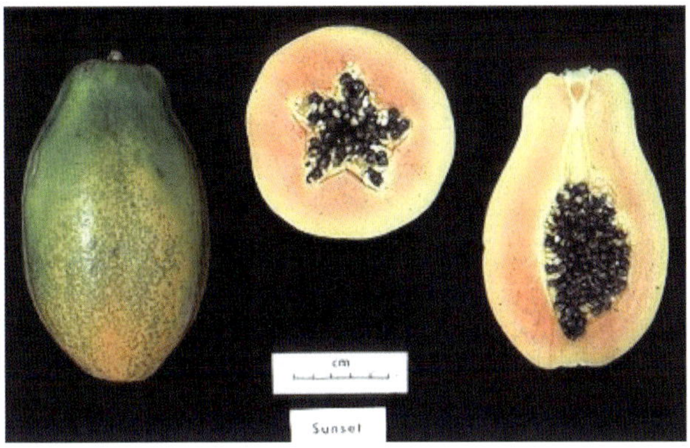

Three views of a papaya, cultivar “Sunset”, which was genetically modified to create the cultivar ‘SunUp’, which is resistant to Papaya ringspot virus.

Papaya was genetically modified to resist the ringspot virus (PSRV). “SunUp” is a transgenic red-fleshed Sunset papaya cultivar that is homozygous for the coat protein gene PRSV; “Rainbow” is a yellow-fleshed F1 hybrid developed by crossing ‘SunUp’ and nontransgenic yellow-fleshed “Kapoho”. *The New York Times* stated, “in the early 1990s, Hawaii’s papaya industry was facing disaster because of the deadly papaya ringspot virus. Its single-handed savior was a breed engineered to be resistant to the virus. Without it, the state’s papaya industry would have collapsed. Today, 80% of Hawaiian papaya is genetically engineered, and there is still no conventional or organic method to control ringspot virus.” The GM cultivar was approved in 1998. In China, a transgenic PRSV-resistant papaya was developed by South China Agricultural University and was first approved for commercial planting in 2006; as of 2012 95% of the papaya grown in Guangdong province and 40% of the papaya grown in Hainan province was genetically modified. In Hong Kong, where there is an exemption on growing and releasing any varieties of GM papaya, more than 80% of grown and imported papayas were transgenic.

The New Leaf potato, a GM food developed using naturally occurring bacteria found in the soil known as *Bacillus thuringiensis* (Bt), was made to provide in-plant protection from the yield-robbing Colorado potato beetle. The New Leaf potato, brought to market by Monsanto in the late 1990s, was developed for the fast food market. It was withdrawn in 2001 after retailers rejected it and food processors ran into export problems.

As of 2005, about 13% of the Zucchini (a form of squash) grown in the US was genetically modified to resist three viruses; that strain is also grown in Canada.

Plums genetically engineered for resistance to plum pox, a disease carried by aphids.

In 2011, BASF requested the European Food Safety Authority's approval for cultivation and marketing of its Fortuna potato as feed and food. The potato was made resistant to late blight by adding resistant genes blb1 and blb2 that originate from the Mexican wild potato Solanum bulbocastanum. In February 2013, BASF withdrew its application.

In 2013, the USDA approved the import of a GM pineapple that is pink in color and that "overexpresses" a gene derived from tangerines and suppress other genes, increasing production of lycopene. The plant's flowering cycle was changed to provide for more uniform growth and quality. The fruit "does not have the ability to propagate and persist in the environment once they have been harvested", according to USDA APHIS. According to Del Monte's submission, the pineapples are commercially grown in a "monoculture" that prevents seed production, as the plant's flowers aren't exposed to compatible pollen sources. Importation into Hawaii is banned for "plant sanitation" reasons.

In 2014, the USDA approved a genetically modified potato developed by J. R. Simplot Company that contained ten genetic modifications that prevent bruising and produce less acrylamide when fried. The modifications eliminate specific proteins from the potatoes, via RNA interference, rather than introducing novel proteins.

In February 2015 Arctic Apples were approved by the USDA, becoming the first genetically modified apple approved for sale in the US. Gene silencing is used to reduce the expression of polyphenol oxidase (PPO), thus preventing the fruit from browning.

Corn

Corn used for food and ethanol has been genetically modified to tolerate various herbicides and to express a protein from *Bacillus thuringiensis* (Bt) that kills certain insects. About 90% of the corn grown in the US was genetically modified in 2010. In the US in 2015, 81% of corn acreage contained the Bt trait and 89% of corn acreage contained the glyphosate-tolerant trait. Corn can be processed into grits, meal and flour as an ingredient in pancakes, muffins, doughnuts, breadings and batters, as well as baby foods, meat products, cereals and some fermented products. Corn-based masa flour and masa dough are used in the production of taco shells, corn chips and tortillas.

Soy

Genetically modified soybean has been modified to tolerate herbicides and produce healthier oils. In 2015, 94% of soybean acreage in the U.S. was genetically modified to be glyphosate-tolerant.

Wheat

As of December 2017, genetically modified wheat has been evaluated in field trials, but has not been released commercially.

Derivative Products

Corn Starch, Starch Sugars and including Syrups

Starch or amylum is a polysaccharide produced by all green plants as an energy store. Pure starch is a white, tasteless and odourless powder. It consists of two types of molecules: the linear and helical amylose and the branched amylopectin. Depending on the plant, starch generally contains 20 to 25% amylose and 75 to 80% amylopectin by weight.

Starch can be further modified to create modified starch for specific purposes, including creation of many of the sugars in processed foods. They include:

- Maltodextrin, a lightly hydrolyzed starch product used as a bland-tasting filler and thickener.
- Various glucose syrups, also called corn syrups in the US, viscous solutions used as sweeteners and thickeners in many kinds of processed foods.
- Dextrose, commercial glucose, prepared by the complete hydrolysis of starch.
- High fructose syrup, made by treating dextrose solutions with the enzyme glucose isomerase, until a substantial fraction of the glucose has been converted to fructose.
- Sugar alcohols, such as maltitol, erythritol, sorbitol, mannitol and hydrogenated starch hydrolysate, are sweeteners made by reducing sugars.

Lecithin

Lecithin is a naturally occurring lipid. It can be found in egg yolks and oil-producing plants. It is an emulsifier and thus is used in many foods. Corn, soy and safflower oil are sources of lecithin, though the majority of lecithin commercially available is derived from soy. Sufficiently processed lecithin is often undetectable with standard testing practices. According to the FDA, no evidence shows or suggests hazard to the public when lecithin is used at common levels. Lecithin added to foods amounts to only

2 to 10 percent of the 1 to 5 g of phosphoglycerides consumed daily on average. Nonetheless, consumer concerns about GM food extend to such products. This concern led to policy and regulatory changes in Europe in 2000, when Regulation (EC) 50/2000 was passed which required labelling of food containing additives derived from GMOs, including lecithin. Because of the difficulty of detecting the origin of derivatives like lecithin with current testing practices, European regulations require those who wish to sell lecithin in Europe to employ a comprehensive system of Identity preservation (IP).

Sugar

The US imports 10% of its sugar, while the remaining 90% is extracted from sugar beet and sugarcane. After deregulation in 2005, glyphosate-resistant sugar beet was extensively adopted in the United States. 95% of beet acres in the US were planted with glyphosate-resistant seed in 2011. GM sugar beets are approved for cultivation in the US, Canada and Japan; the vast majority are grown in the US. GM beets are approved for import and consumption in Australia, Canada, Colombia, EU, Japan, Korea, Mexico, New Zealand, Philippines, the Russian Federation and Singapore. Pulp from the refining process is used as animal feed. The sugar produced from GM sugar beets contains no DNA or protein – it is just sucrose that is chemically indistinguishable from sugar produced from non-GM sugar beets. Independent analyses conducted by internationally recognized laboratories found that sugar from Roundup Ready sugar beets is identical to the sugar from comparably grown conventional (non-Roundup Ready) sugar beets.

Vegetable Oil

Most vegetable oil used in the US is produced from GM crops canola, corn, cotton and soybeans. Vegetable oil is sold directly to consumers as cooking oil, shortening and margarine and is used in prepared foods. There is a vanishingly small amount of protein or DNA from the original crop in vegetable oil. Vegetable oil is made of triglycerides extracted from plants or seeds and then refined and may be further processed via hydrogenation to turn liquid oils into solids. The refining process removes all, or nearly all non-triglyceride ingredients. Medium-chain triglycerides (MCTs) offer an alternative to conventional fats and oils. The length of a fatty acid influences its fat absorption during the digestive process. Fatty acids in the middle position on the glycerol molecules appear to be absorbed more easily and influence metabolism more than fatty acids on the end positions. Unlike ordinary fats, MCTs are metabolized like carbohydrates. They have exceptional oxidative stability, and prevent foods from turning rancid readily.

Animal Feed

Livestock and poultry are raised on animal feed, much of which is composed of the leftovers from processing crops, including GM crops. For example, approximately 43%

of a canola seed is oil. What remains after oil extraction is a meal that becomes an ingredient in animal feed and contains canola protein. Likewise, the bulk of the soybean crop is grown for oil and meal. The high-protein defatted and toasted soy meal becomes livestock feed and dog food. 98% of the US soybean crop goes for livestock feed. In 2011, 49% of the US maize harvest was used for livestock feed (including the percentage of waste from distillers grains). "Despite methods that are becoming more and more sensitive, tests have not yet been able to establish a difference in the meat, milk, or eggs of animals depending on the type of feed they are fed. It is impossible to tell if an animal was fed GM soy just by looking at the resulting meat, dairy, or egg products. The only way to verify the presence of GMOs in animal feed is to analyze the origin of the feed itself."

A 2012 literature review of studies evaluating the effect of GM feed on the health of animals did not find evidence that animals were adversely affected, although small biological differences were occasionally found. The studies included in the review ranged from 90 days to two years, with several of the longer studies considering reproductive and intergenerational effects.

Enzymes produced by genetically modified microorganisms are also integrated into animal feed to enhance availability of nutrients and overall digestion. These enzymes may also provide benefit to the gut microbiome of an animal, as well as hydrolyse antinutritional factors present in the feed.

Proteins

Rennet is a mixture of enzymes used to coagulate milk into cheese. Originally it was available only from the fourth stomach of calves, and was scarce and expensive, or was available from microbial sources, which often produced unpleasant tastes. Genetic engineering made it possible to extract rennet-producing genes from animal stomachs and insert them into bacteria, fungi or yeasts to make them produce chymosin, the key enzyme. The modified microorganism is killed after fermentation. Chymosin is isolated from the fermentation broth, so that the Fermentation-Produced Chymosin (FPC) used by cheese producers has an amino acid sequence that is identical to bovine rennet. The majority of the applied chymosin is retained in the whey. Trace quantities of chymosin may remain in cheese.

FPC was the first artificially produced enzyme to be approved by the US Food and Drug Administration. FPC products have been on the market since 1990 and as of 2015 had yet to be surpassed in commercial markets. In 1999, about 60% of US hard cheese was made with FPC. Its global market share approached 80%. By 2008, approximately 80% to 90% of commercially made cheeses in the US and Britain were made using FPC.

In some countries, recombinant (GM) bovine somatotropin (also called rBST, or bovine growth hormone or BGH) is approved for administration to increase milk production.

rBST may be present in milk from rBST treated cows, but it is destroyed in the digestive system and even if directly injected into the human bloodstream, has no observable effect on humans. The FDA, World Health Organization, American Medical Association, American Dietetic Association and the National Institutes of Health have independently stated that dairy products and meat from rBST-treated cows are safe for human consumption. However, on 30 September 2010, the United States Court of Appeals, Sixth Circuit, analyzing submitted evidence, found a "compositional difference" between milk from rBGH-treated cows and milk from untreated cows. The court stated that milk from rBGH-treated cows has: increased levels of the hormone Insulin-like growth factor 1 (IGF-1); higher fat content and lower protein content when produced at certain points in the cow's lactation cycle; and more somatic cell counts, which may "make the milk turn sour more quickly".

Livestock

Genetically modified livestock are organisms from the group of cattle, sheep, pigs, goats, birds, horses and fish kept for human consumption, whose genetic material (DNA) has been altered using genetic engineering techniques. In some cases, the aim is to introduce a new trait to the animals which does not occur naturally in the species, i.e. transgenesis.

A 2003 review published on behalf of Food Standards Australia New Zealand examined transgenic experimentation on terrestrial livestock species as well as aquatic species such as fish and shellfish. The review examined the molecular techniques used for experimentation as well as techniques for tracing the transgenes in animals and products as well as issues regarding transgene stability.

Some mammals typically used for food production have been modified to produce non-food products, a practice sometimes called Pharming.

Salmon

A GM salmon, awaiting regulatory approval since 1997, was approved for human consumption by the American FDA in November 2015, to be raised in specific land-based hatcheries in Canada and Panama.

Health and Safety

There is a scientific consensus that currently available food derived from GM crops poses no greater risk to human health than conventional food, but that each GM food needs to be tested on a case-by-case basis before introduction. Nonetheless, members of the public are much less likely than scientists to perceive GM foods as safe. The legal and regulatory status of GM foods varies by country, with some nations banning or restricting them, and others permitting them with widely differing degrees of regulation.

Opponents claim that long-term health risks have not been adequately assessed and propose various combinations of additional testing, labeling or removal from the market. The advocacy group European Network of Scientists for Social and Environmental Responsibility (ENSSER), disputes the claim that "science" supports the safety of current GM foods, proposing that each GM food must be judged on case-by-case basis.

Testing

The legal and regulatory status of GM foods varies by country, with some nations banning or restricting them, and others permitting them with widely differing degrees of regulation. Countries such as the United States, Canada, Lebanon and Egypt use *substantial equivalence* to determine if further testing is required, while many countries such as those in the European Union, Brazil and China only authorize GMO cultivation on a case-by-case basis. In the U.S. the FDA determined that GMO's are "Generally Recognized as Safe" (GRAS) and therefore do not require additional testing if the GMO product is substantially equivalent to the non-modified product. If new substances are found, further testing may be required to satisfy concerns over potential toxicity, allergenicity, possible gene transfer to humans or genetic outcrossing to other organisms.

Genetically Modified Crops

A GM or transgenic crop is a plant that has a novel combination of genetic material obtained through the use of modern biotechnology.

For example, a GM crop can contain a gene(s) that has been artificially inserted instead of the plant acquiring it through pollination.

The resulting plant is said to be "genetically modified" although in reality all crops have been "genetically modified" from their original wild state by domestication, selection, and controlled breeding over long periods of time.

In the developed world, there is clear evidence that the use of GM crops has resulted in significant benefits. These include:

1. Higher crop yields.
2. Reduced farm costs.
3. Increased farm profit.
4. Safer environment.
5. More nutritious food.

The "first generation" crops with traits such as insect resistance and herbicide tolerance have proven their ability to lower farm-level production costs.

The "second-generation" GM crops feature increased nutritional and/or industrial traits. These crops have more direct benefits to consumers. Examples of commercialized second generation crops include (ISAAA GM Approval Database):

1. Non-browning apples.
2. Non-bruising and low acrylamide potatoes.
3. Maize varieties with low phytic acid and increased essential amino acids.
4. Healthier oils from soybean and canola.

Methods

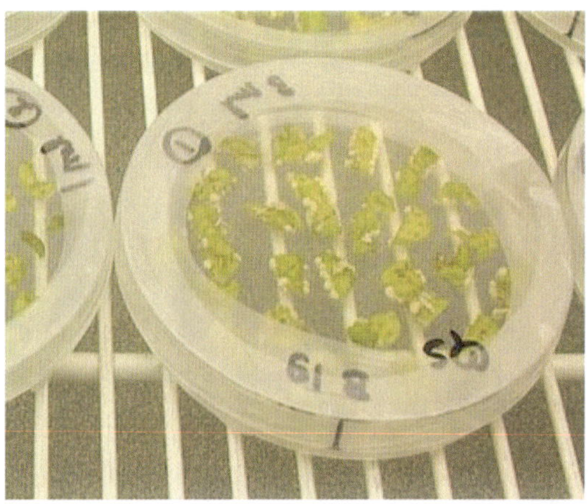

Plants (Solanum chacoense) being transformed using agrobacterium.

Genetically engineered crops have genes added or removed using genetic engineering techniques, originally including gene guns, electroporation, microinjection and agrobacterium. More recently, CRISPR and TALEN offered much more precise and convenient editing techniques.

Gene guns (also known as biolistics) "shoot" (direct high energy particles or radiations against) target genes into plant cells. It is the most common method. DNA is bound to tiny particles of gold or tungsten which are subsequently shot into plant tissue or single plant cells under high pressure. The accelerated particles penetrate both the cell wall and membranes. The DNA separates from the metal and is integrated into plant DNA inside the nucleus. This method has been applied successfully for many cultivated crops, especially monocots like wheat or maize, for which transformation using *Agrobacterium tumefaciens* has been less successful. The major disadvantage of this procedure is that serious damage can be done to the cellular tissue.

Agrobacterium tumefaciens-mediated transformation is another common technique. Agrobacteria are natural plant parasites. Their natural ability to transfer genes provides another engineering method. To create a suitable environment for themselves, these Agrobacteria insert their genes into plant hosts, resulting in a proliferation of modified plant cells near the soil level (crown gall). The genetic information for tumor growth is encoded on a mobile, circular DNA fragment (plasmid). When *Agrobacterium* infects a plant, it transfers this T-DNA to a random site in the plant genome. When used in genetic engineering the bacterial T-DNA is removed from the bacterial plasmid and replaced with the desired foreign gene. The bacterium is a vector, enabling transportation of foreign genes into plants. This method works especially well for dicotyledonous plants like potatoes, tomatoes, and tobacco. Agrobacteria infection is less successful in crops like wheat and maize.

Electroporation is used when the plant tissue does not contain cell walls. In this technique, "DNA enters the plant cells through miniature pores which are temporarily caused by electric pulses." Microinjection is used to directly inject foreign DNA into cells. Plant scientists, backed by results of modern comprehensive profiling of crop composition, point out that crops modified using GM techniques are less likely to have unintended changes than are conventionally bred crops.

In research tobacco and *Arabidopsis thaliana* are the most frequently modified plants, due to well-developed transformation methods, easy propagation and well studied genomes. They serve as model organisms for other plant species. Introducing new genes into plants requires a promoter specific to the area where the gene is to be expressed. For instance, to express a gene only in rice grains and not in leaves, an endosperm-specific promoter is used. The codons of the gene must be optimized for the organism due to codon usage bias.

Types of Modifications

Transgenic maize containing a gene from the bacteria *Bacillus thuringiensis*.

Transgenic

Transgenic plants have genes inserted into them that are derived from another species. The inserted genes can come from species within the same kingdom (plant to plant), or between kingdoms (for example, bacteria to plant). In many cases the inserted DNA has to be modified slightly in order to be correctly and efficiently expressed in the host organism. Transgenic plants are used to express proteins, like the cry toxins from *B. thuringiensis*, herbicide-resistant genes, antibodies, and antigens for vaccinations. A study led by the European Food Safety Authority (EFSA) also found viral genes in transgenic plants.

Transgenic carrots have been used to produce the drug Taliglucerase alfa which is used to treat Gaucher's disease. In the laboratory, transgenic plants have been modified to increase photosynthesis (currently about 2% at most plants versus the theoretic potential of 9–10%). This is possible by changing the rubisco enzyme (i.e. changing C_3 plants into C_4 plants), by placing the rubisco in a carboxysome, by adding CO_2 pumps in the cell wall, or by changing the leaf form or size. Plants have been engineered to exhibit bioluminescence that may become a sustainable alternative to electric lighting.

Cisgenic

Cisgenic plants are made using genes found within the same species or a closely related one, where conventional plant breeding can occur. Some breeders and scientists argue that cisgenic modification is useful for plants that are difficult to crossbreed by conventional means (such as potatoes), and that plants in the cisgenic category should not require the same regulatory scrutiny as transgenics.

Subgenic

Genetically modified plants can also be developed using gene knockdown or gene knockout to alter the genetic makeup of a plant without incorporating genes from other plants. In 2014, Chinese researcher Gao Caixia filed patents on the creation of a strain of wheat that is resistant to powdery mildew. The strain lacks genes that encode proteins that repress defenses against the mildew. The researchers deleted all three copies of the genes from wheat's hexaploid genome. Gao used the TALENs and CRISPR gene editing tools without adding or changing any other genes. No field trials were immediately planned. The CRISPR technique has also been used by Penn State researcher Yinong Yang to modify white button mushrooms (*Agaricus bisporus*) to be non-browning, and by DuPont Pioneer to make a new variety of corn.

Multiple Trait Integration

With multiple trait integration, several new traits may be integrated into a new crop.

Economics

GM food's economic value to farmers is one of its major benefits, including in developing nations. A 2010 study found that Bt corn provided economic benefits of $6.9 billion over the previous 14 years in five Midwestern states. The majority ($4.3 billion) accrued to farmers producing non-Bt corn. This was attributed to European corn borer populations reduced by exposure to Bt corn, leaving fewer to attack conventional corn nearby. Agriculture economists calculated that "world surplus [increased by] $240.3 million for 1996. Of this total, the largest share (59%) went to U.S. farmers. Seed company Monsanto received the next largest share (21%), followed by US consumers (9%), the rest of the world (6%), and the germplasm supplier, Delta & Pine Land Company of Mississippi (5%)."

According to the International Service for the Acquisition of Agri-biotech Applications (ISAAA), in 2014 approximately 18 million farmers grew biotech crops in 28 countries; about 94% of the farmers were resource-poor in developing countries. 53% of the global biotech crop area of 181.5 million hectares was grown in 20 developing countries. PG Economics comprehensive 2012 study concluded that GM crops increased farm incomes worldwide by $14 billion in 2010, with over half this total going to farmers in developing countries.

Critics challenged the claimed benefits to farmers over the prevalence of biased observers and by the absence of randomized controlled trials. The main Bt crop grown by small farmers in developing countries is cotton. A 2006 review of Bt cotton findings by agricultural economists concluded, "the overall balance sheet, though promising, is mixed. Economic returns are highly variable over years, farm type, and geographical location".

In 2013 the European Academies Science Advisory Council (EASAC) asked the EU to allow the development of agricultural GM technologies to enable more sustainable agriculture, by employing fewer land, water, and nutrient resources. EASAC also criticizes the EU's "time-consuming and expensive regulatory framework" and said that the EU had fallen behind in the adoption of GM technologies.

Participants in agriculture business markets include seed companies, agrochemical companies, distributors, farmers, grain elevators and universities that develop new crops/traits and whose agricultural extensions advise farmers on best practices. According to a 2012 review based on data from the late 1990s and early 2000s, much of the GM crop grown each year is used for livestock feed and increased demand for meat leads to increased demand for GM feed crops. Feed grain usage as a percentage of total crop production is 70% for corn and more than 90% of oil seed meals such as soybeans. About 65 million metric tons of GM corn grains and about 70 million metric tons of soybean meals derived from GM soybean become feed.

In 2014 the global value of biotech seed was US$15.7 billion; US$11.3 billion (72%) was in industrial countries and US$4.4 billion (28%) was in the developing countries. In

2009, Monsanto had $7.3 billion in sales of seeds and from licensing its technology; DuPont, through its Pioneer subsidiary, was the next biggest company in that market. As of 2009, the overall Roundup line of products including the GM seeds represented about 50% of Monsanto's business.

Some patents on GM traits have expired, allowing the legal development of generic strains that include these traits. For example, generic glyphosate-tolerant GM soybean is now available. Another impact is that traits developed by one vendor can be added to another vendor's proprietary strains, potentially increasing product choice and competition. The patent on the first type of *Roundup Ready* crop that Monsanto produced (soybeans) expired in 2014 and the first harvest of off-patent soybeans occurs in the spring of 2015. Monsanto has broadly licensed the patent to other seed companies that include the glyphosate resistance trait in their seed products. About 150 companies have licensed the technology, including Syngenta and DuPont Pioneer.

Yield

In 2014, the largest review yet concluded that GM crops' effects on farming were positive. The meta-analysis considered all published English-language examinations of the agronomic and economic impacts between 1995 and March 2014 for three major GM crops: soybean, maize, and cotton. The study found that herbicide-tolerant crops have lower production costs, while for insect-resistant crops the reduced pesticide use was offset by higher seed prices, leaving overall production costs about the same.

Yields increased 9% for herbicide tolerance and 25% for insect resistant varieties. Farmers who adopted GM crops made 69% higher profits than those who did not. The review found that GM crops help farmers in developing countries, increasing yields by 14 percentage points.

The researchers considered some studies that were not peer-reviewed and a few that did not report sample sizes. The large data set allowed the study to control for potentially confounding variables such as fertilizer use. Separately, they concluded that the funding source did not influence study results.

Traits

GM crops grown today, or under development, have been modified with various traits. These traits include improved shelf life, disease resistance, stress resistance, herbicide resistance, pest resistance, production of useful goods such as biofuel or drugs, and ability to absorb toxins and for use in bioremediation of pollution.

Recently, research and development has been targeted to enhancement of crops that are locally important in developing countries, such as insect-resistant cowpea for Africa and insect-resistant brinjal (eggplant).

Extended Shelf Life

The first genetically modified crop approved for sale in the U.S. was the *FlavrSavr* tomato, which had a longer shelf life. First sold in 1994, FlavrSavr tomato production ceased in 1997. It is no longer on the market.

In November 2014, the USDA approved a GM potato that prevents bruising.

In February 2015 Arctic Apples were approved by the USDA, becoming the first genetically modified apple approved for US sale. Gene silencing was used to reduce the expression of polyphenol oxidase (PPO), thus preventing enzymatic browning of the fruit after it has been sliced open. The trait was added to Granny Smith and Golden Delicious varieties. The trait includes a bacterial antibiotic resistance gene that provides resistance to the antibiotic kanamycin. The genetic engineering involved cultivation in the presence of kanamycin, which allowed only resistant cultivars to survive. Humans consuming apples do not acquire kanamycin resistance, per arcticapple.com. The FDA approved the apples in March 2015.

Improved Photosynthesis

Plants use non-photochemical quenching (NPQ) to protect them from excessive amounts of sunlight. Plants can switch on the quenching mechanism almost instantaneously but it takes much longer for it to switch off again. During the time that it is switched off, the amount of energy that is wasted increases. A genetic modification in three genes allows to correct this (in a trial with tobacco plants). As a result, yields were 14-20% higher, in terms of the weight of the dry leaves harvested. The plants had bigger leaves, were taller and had more vigorous roots.

Another improvement that can be made on the photosynthesis process (with C3 pathway plants) is on photorespiration. By inserting the C4 pathway into C3 plants, productivity may increase by as much as 50% for cereal crops, such as rice.

Improved Nutritional Value

Edible Oils

Some GM soybeans offer improved oil profiles for processing. Camelina sativa has been modified to produce plants that accumulate high levels of oils similar to fish oils.

Vitamin Enrichment

Golden rice, developed by the International Rice Research Institute (IRRI), provides greater amounts of vitamin A targeted at reducing vitamin A deficiency. As of January 2016, golden rice has not yet been grown commercially in any country.

Toxin Reduction

A genetically modified cassava under development offers lower cyanogen glucosides and enhanced protein and other nutrients (called BioCassava).

In November 2014, the USDA approved a potato that prevents bruising and produces less acrylamide when fried. They do not employ genes from non-potato species. The trait was added to the Russet Burbank, Ranger Russet and Atlantic varieties.

Stress Resistance

Plants have been engineered to tolerate non-biological stressors, such as drought, frost, and high soil salinity. In 2011, Monsanto's DroughtGard maize became the first drought-resistant GM crop to receive US marketing approval.

Drought resistance occurs by modifying the plant's genes responsible for the mechanism known as the crassulacean acid metabolism (CAM), which allows the plants to survive despite low water levels. This holds promise for water-heavy crops such as rice, wheat, soybeans and poplar to accelerate their adaptation to water-limited environments. Several salinity tolerance mechanisms have been identified in salt-tolerant crops. For example, rice, canola and tomato crops have been genetically modified to increase their tolerance to salt stress.

Herbicides

Glyphosate

As of 1999, the most prevalent GM trait was glyphosate-tolerance. Glyphosate (the active ingredient in Roundup and other herbicide products) kills plants by interfering with the shikimate pathway in plants, which is essential for the synthesis of the aromatic amino acids phenylalanine, tyrosine, and tryptophan. The shikimate pathway is not present in animals, which instead obtain aromatic amino acids from their diet. More specifically, glyphosate inhibits the enzyme 5-enolpyruvylshikimate-3-phosphate synthase (EPSPS).

This trait was developed because the herbicides used on grain and grass crops at the time were highly toxic and not effective against narrow-leaved weeds. Thus, developing crops that could withstand spraying with glyphosate would both reduce environmental and health risks, and give an agricultural edge to the farmer.

Some micro-organisms have a version of EPSPS that is resistant to glyphosate inhibition. One of these was isolated from an *Agrobacterium* strain CP4 (CP4 EPSPS) that was resistant to glyphosate. The CP4 EPSPS gene was engineered for plant expression by fusing the 5' end of the gene to a chloroplast transit peptide derived from the petunia EPSPS. This transit peptide was used because it had shown previously an ability to deliver bacterial EPSPS to the chloroplasts of other plants. This CP4 EPSPS gene was cloned and transfected into soybeans.

The plasmid used to move the gene into soybeans was PV-GMGTO4. It contained three bacterial genes, two CP4 EPSPS genes, and a gene encoding beta-glucuronidase (GUS) from *Escherichia coli* as a marker. The DNA was injected into the soybeans using the particle acceleration method. Soybean cultivar A5403 was used for the transformation.

Bromoxynil

Tobacco plants have been engineered to be resistant to the herbicide bromoxynil.

Glufosinate

Crops have been commercialized that are resistant to the herbicide glufosinate, as well. Crops engineered for resistance to multiple herbicides to allow farmers to use a mixed group of two, three, or four different chemicals are under development to combat growing herbicide resistance.

2,4-D

In October 2014 the US EPA registered Dow's Enlist Duo maize, which is genetically modified to be resistant to both glyphosate and 2,4-D, in six states. Inserting a bacterial aryloxyalkanoate dioxygenase gene, *aad1* makes the corn resistant to 2,4-D. The USDA had approved maize and soybeans with the mutation in September 2014.

Dicamba

Monsanto has requested approval for a stacked strain that is tolerant of both glyphosate and dicamba. The request includes plans for avoiding herbicide drift to other crops. Significant damage to other non-resistant crops occurred from dicamba formulations intended to reduce volatilization drifting when sprayed on resistant soybeans in 2017. The newer dicamba formulation labels specify to not spray when average wind speeds are above 10-15 miles per hour (16-24 km/h) to avoid particle drift, average wind speeds below 3 miles per hour (4.8 km/h) to avoid temperature inversions, and rain or high temperatures are in the next day forecast. However, these conditions typically only occur during June and July for a few hours at a time.

Pest Resistance

Insects

Tobacco, corn, rice and some other crops have been engineered to express genes encoding for insecticidal proteins from Bacillus thuringiensis (Bt). The introduction of Bt crops during the period between 1996 and 2005 has been estimated to have reduced the total volume of insecticide active ingredient use in the United States by over 100 thousand tons. This represents a 19.4% reduction in insecticide use. In the late 1990s, a genetically modified potato that was resistant to the Colorado potato beetle was withdrawn because major buyers rejected it, fearing consumer opposition.

Viruses

Papaya, potatoes, and squash have been engineered to resist viral pathogens such as cucumber mosaic virus which, despite its name, infects a wide variety of plants. Virus resistant papaya were developed in response to a papaya ringspot virus (PRV) outbreak in Hawaii in the late 1990s. They incorporate PRV DNA. By 2010, 80% of Hawaiian papaya plants were genetically modified.

Potatoes were engineered for resistance to potato leaf roll virus and Potato virus Y in 1998. Poor sales led to their market withdrawal after three years. Yellow squash that were resistant to at first two, then three viruses were developed, beginning in the 1990s. The viruses are watermelon, cucumber and zucchini/courgette yellow mosaic. Squash was the second GM crop to be approved by US regulators. The trait was later added to zucchini.

Many strains of corn have been developed in recent years to combat the spread of Maize dwarf mosaic virus, a costly virus that causes stunted growth which is carried in Johnson grass and spread by aphid insect vectors. These strands are commercially available although the resistance is not standard among GM corn variants.

By-products

Drugs

In 2012, the FDA approved the first plant-produced pharmaceutical, a treatment for Gaucher's Disease. Tobacco plants have been modified to produce therapeutic antibodies.

Biofuel

Algae is under development for use in biofuels. Researchers in Singapore were working on GM jatropha for biofuel production. Syngenta has USDA approval to market a maize trademarked Enogen that has been genetically modified to convert its starch to sugar for ethanol. Some trees have been genetically modified to either have less lignin, or to express lignin with chemically labile bonds. Lignin is the critical limiting factor when using wood to make bio-ethanol because lignin limits the accessibility of cellulose microfibrils to depolymerization by enzymes. Besides with trees, the chemically labile lignin bonds are also very useful for cereal crops such as maize, barley, and oats.

Materials

Companies and labs are working on plants that can be used to make bioplastics. Potatoes that produce industrially useful starches have been developed as well. Oilseed can be modified to produce fatty acids for detergents, substitute fuels and petrochemicals.

Bioremediation

Scientists at the University of York developed a weed (*Arabidopsis thaliana*) that contains genes from bacteria that could clean TNT and RDX-explosive soil contaminants in 2011. 16 million hectares in the US (1.5% of the total surface) are estimated to be contaminated with TNT and RDX. However *A. thaliana* was not tough enough for use on military test grounds. Modifications in 2016 included switchgrass and bentgrass.

Genetically modified plants have been used for bioremediation of contaminated soils. Mercury, selenium and organic pollutants such as polychlorinated biphenyls (PCBs).

Marine environments are especially vulnerable since pollution such as oil spills are not containable. In addition to anthropogenic pollution, millions of tons of petroleum annually enter the marine environment from natural seepages. Despite its toxicity, a considerable fraction of petroleum oil entering marine systems is eliminated by the hydrocarbon-degrading activities of microbial communities. Particularly successful is a recently discovered group of specialists, the so-called hydrocarbonoclastic bacteria (HCCB) that may offer useful genes.

Asexual Reproduction

Crops such as maize reproduce sexually each year. This randomizes which genes get propagated to the next generation, meaning that desirable traits can be lost. To maintain a high-quality crop, some farmers purchase seeds every year. Typically, the seed company maintains two inbred varieties and crosses them into a hybrid strain that is then sold. Related plants like sorghum and gamma grass are able to perform apomixis, a form of asexual reproduction that keeps the plant's DNA intact. This trait is apparently controlled by a single dominant gene, but traditional breeding has been unsuccessful in creating asexually-reproducing maize. Genetic engineering offers another route to this goal. Successful modification would allow farmers to replant harvested seeds that retain desirable traits, rather than relying on purchased seed.

Genetic modifications to some crops also exist, which make it easier to process the crop. Also, some crops (such as tomatoes) have been genetic modified to contain no seed at all.

Farming Practices

Bacillus Thuringiensis

Constant exposure to a toxin creates evolutionary pressure for pests resistant to that toxin. Over-reliance on glyphosate and a reduction in the diversity of weed management practices allowed the spread of glyphosate resistance in 14 weed species in the US, and in soybeans.

To reduce resistance to *Bacillus thuringiensis* (Bt) crops, the 1996 commercialization of transgenic cotton and maize came with a management strategy to prevent insects from becoming resistant. Insect resistance management plans are mandatory for Bt crops. The aim is to encourage a large population of pests so that any (recessive) resistance genes are diluted within the population. Resistance lowers evolutionary fitness in the absence of the stressor, Bt. In refuges, non-resistant strains outcompete resistant ones.

With sufficiently high levels of transgene expression, nearly all of the heterozygotes (S/s), i.e., the largest segment of the pest population carrying a resistance allele, will be killed before maturation, thus preventing transmission of the resistance gene to their progeny. Refuges (i. e., fields of nontransgenic plants) adjacent to transgenic fields increases the likelihood that homozygous resistant (s/s) individuals and any surviving heterozygotes will mate with susceptible (S/S) individuals from the refuge, instead of with other individuals carrying the resistance allele. As a result, the resistance gene frequency in the population remains lower.

Complicating factors can affect the success of the high-dose/refuge strategy. For example, if the temperature is not ideal, thermal stress can lower Bt toxin production and leave the plant more susceptible. More importantly, reduced late-season expression has been documented, possibly resulting from DNA methylation of the promoter. The success of the high-dose/refuge strategy has successfully maintained the value of Bt crops. This success has depended on factors independent of management strategy, including low initial resistance allele frequencies, fitness costs associated with resistance, and the abundance of non-Bt host plants outside the refuges.

Companies that produce Bt seed are introducing strains with multiple Bt proteins. Monsanto did this with Bt cotton in India, where the product was rapidly adopted. Monsanto has also; in an attempt to simplify the process of implementing refuges in fields to comply with Insect Resistance Management (IRM) policies and prevent irresponsible planting practices; begun marketing seed bags with a set proportion of refuge (non-transgenic) seeds mixed in with the Bt seeds being sold. Coined "Refuge-In-a-Bag" (RIB), this practice is intended to increase farmer compliance with refuge requirements and reduce additional labor needed at planting from having separate Bt and refuge seed bags on hand. This strategy is likely to reduce the likelihood of Bt-resistance occurring for corn rootworm, but may increase the risk of resistance for lepidopteran corn pests, such as European corn borer. Increased concerns for resistance with seed mixtures include partially resistant larvae on a Bt plant being able to move to a susceptible plant to survive or cross pollination of refuge pollen on to Bt plants that can lower the amount of Bt expressed in kernels for ear feeding insects.

Herbicide Resistance

Best management practices (BMPs) to control weeds may help delay resistance. BMPs include applying multiple herbicides with different modes of action, rotating crops, planting weed-free seed, scouting fields routinely, cleaning equipment to reduce the transmission of weeds to other fields, and maintaining field borders. The most widely planted GM crops are designed to tolerate herbicides. By 2006 some weed populations had evolved to tolerate some of the same herbicides. Palmer amaranth is a weed that competes with cotton. A native of the southwestern US, it traveled east and was first found resistant to glyphosate in 2006, less than 10 years after GM cotton was introduced.

Plant Protection

Farmers generally use less insecticide when they plant Bt-resistant crops. Insecticide use on corn farms declined from 0.21 pound per planted acre in 1995 to 0.02 pound in 2010. This is consistent with the decline in European corn borer populations as a direct result of Bt corn and cotton. The establishment of minimum refuge requirements helped delay the evolution of Bt resistance. However, resistance appears to be developing to some Bt traits in some areas.

Tillage

By leaving at least 30% of crop residue on the soil surface from harvest through planting, conservation tillage reduces soil erosion from wind and water, increases water retention, and reduces soil degradation as well as water and chemical runoff. In addition, conservation tillage reduces the carbon footprint of agriculture. A 2014 review covering 12 states from 1996 to 2006, found that a 1% increase in herbicde-tolerant (HT) soybean adoption leads to a 0.21% increase in conservation tillage and a 0.3% decrease in quality-adjusted herbicide use.

Regulation

The regulation of genetic engineering concerns the approaches taken by governments to assess and manage the risks associated with the development and release of genetically modified crops. There are differences in the regulation of GM crops between countries, with some of the most marked differences occurring between the US and Europe. Regulation varies in a given country depending on the intended use of each product. For example, a crop not intended for food use is generally not reviewed by authorities responsible for food safety.

Production

In 2013, GM crops were planted in 27 countries; 19 were developing countries and 8 were developed countries. 2013 was the second year in which developing countries grew a majority (54%) of the total GM harvest. 18 million farmers grew GM crops; around 90% were small-holding farmers in developing countries.

Country	2013– GM planted area (million hectares)	Biotech crops
US	70.1	Maize, Soybean, Cotton, Canola, Sugar-beet, Alfalfa, Papaya, Squash
Brazil	40.3	Soybean, Maize, Cotton
Argentina	24.4	Soybean, Maize, Cotton
India	11.0	Cotton
Canada	10.8	Canola, Maize, Soybean, Sugarbeet
Total	175.2	----

The United States Department of Agriculture (USDA) reports every year on the total area of GM crop varieties planted in the United States. According to National Agricultural Statistics Service, the states published in these tables represent 81–86 percent of all corn planted area, 88–90 percent of all soybean planted area, and 81–93 percent of all upland cotton planted area (depending on the year).

Global estimates are produced by the International Service for the Acquisition of Agri-biotech Applications (ISAAA) and can be found in their annual reports, "Global Status of Commercialized Transgenic Crops".

Farmers have widely adopted GM technology. Between 1996 and 2013, the total surface area of land cultivated with GM crops increased by a factor of 100, from 17,000 square kilometers (4,200,000 acres) to 1,750,000 km^2 (432 million acres). 10% of the world's arable land was planted with GM crops in 2010. As of 2011, 11 different transgenic crops were grown commercially on 395 million acres (160 million hectares) in 29 countries such as the US, Brazil, Argentina, India, Canada, China, Paraguay, Pakistan, South Africa, Uruguay, Bolivia, Australia, Philippines, Myanmar, Burkina Faso, Mexico and Spain. One of the key reasons for this widespread adoption is the perceived economic benefit the technology brings to farmers. For example, the system of planting glyphosate-resistant seed and then applying glyphosate once plants emerged provided farmers with the opportunity to dramatically increase the yield from a given plot of land, since this allowed them to plant rows closer together. Without it, farmers had to plant rows far enough apart to control post-emergent weeds with mechanical tillage. Likewise, using Bt seeds means that farmers do not have to purchase insecticides, and then invest time, fuel, and equipment in applying them. However critics have disputed whether yields are higher and whether chemical use is less, with GM crops.

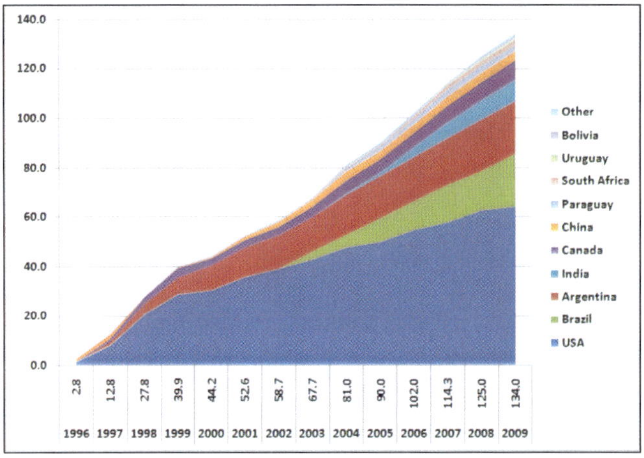

Land area used for genetically modified crops by country, in millions of hectares.
In 2011, the land area used was 160 million hectares, or 1.6 million square kilometers.

In the US, by 2014, 94% of the planted area of soybeans, 96% of cotton and 93% of corn were genetically modified varieties. Genetically modified soybeans carried herbicide-tolerant traits only, but maize and cotton carried both herbicide tolerance and insect protection traits (the latter largely Bt protein). These constitute "input-traits" that are aimed to financially benefit the producers, but may have indirect environmental benefits and cost benefits to consumers. The Grocery Manufacturers of America estimated in 2003 that 70–75% of all processed foods in the U.S. contained a GM ingredient.

Europe grows relatively few genetically engineered crops with the exception of Spain, where one fifth of maize is genetically engineered, and smaller amounts in five other countries. The EU had a 'de facto' ban on the approval of new GM crops, from 1999 until 2004. GM crops are now regulated by the EU. In 2015, genetically engineered crops are banned in 38 countries worldwide, 19 of them in Europe. Developing countries grew 54 percent of genetically engineered crops in 2013.

In recent years GM crops expanded rapidly in developing countries. In 2013 approximately 18 million farmers grew 54% of worldwide GM crops in developing countries. 2013's largest increase was in Brazil (403,000 km^2 versus 368,000 km^2 in 2012). GM cotton began growing in India in 2002, reaching 110,000 km^2 in 2013.

According to the 2013 ISAAA brief: "a total of 36 countries (35 + EU-28) have granted regulatory approvals for biotech crops for food and/or feed use and for environmental release or planting since 1994 a total of 2,833 regulatory approvals involving 27 GM crops and 336 GM events (NB: an "event" is a specific genetic modification in a specific species) have been issued by authorities, of which 1,321 are for food use (direct use or processing), 918 for feed use (direct use or processing) and 599 for environmental release or planting. Japan has the largest number (198), followed by the U.S.A. (165, not including "stacked" events), Canada (146), Mexico (131), South Korea (103), Australia

(93), New Zealand (83), European Union (71 including approvals that have expired or under renewal process), Philippines (68), Taiwan (65), Colombia (59), China (55) and South Africa (52). Maize has the largest number (130 events in 27 countries), followed by cotton (49 events in 22 countries), potato (31 events in 10 countries), canola (30 events in 12 countries) and soybean (27 events in 26 countries).

Gene Therapy

Gene therapy is an experimental technique that uses genes to treat or prevent disease. The new DNA usually contains a functioning gene to correct the effects of a disease-causing mutation.

Gene therapy uses sections of DNA(usually genes) to treat or prevent disease. The DNA is carefully selected to correct the effect of a mutated gene that is causing disease. The technique was first developed in 1972 but has, so far, had limited success in treating human diseases. Gene therapy may be a promising treatment option for some genetic diseases, including muscular dystrophy and cystic fibrosis.

There are two different types of gene therapy depending on which types of cells are treated:

- Somatic gene therapy: transfer of a section of DNA to any cell of the body that doesn't produce sperm or eggs. Effects of gene therapy will not be passed onto the patient's children.
- Germline gene therapy: transfer of a section of DNA to cells that produce eggs or sperm. Effects of gene therapy will be passed onto the patient's children and subsequent generations.

Gene Therapy Techniques

There are several techniques for carrying out gene therapy. These include:

Gene Augmentation Therapy

This is used to treat diseases caused by a mutation that stops a gene from producing a functioning product, such as a protein. This therapy adds DNA containing a functional version of the lost gene back into the cell. The new gene produces a functioning product at sufficient levels to replace the protein that was originally missing. This is only successful if the effects of the disease are reversible or have not resulted in lasting damage to the body. For example, this can be used to treat loss of function disorders such as cystic fibrosis by introducing a functional copy of the gene to correct the disease.

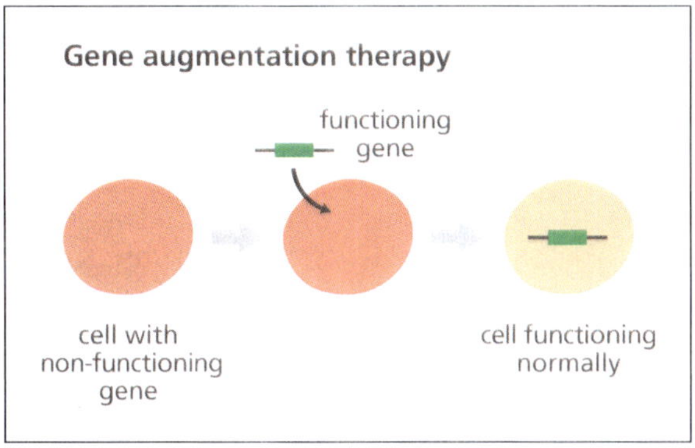

Gene inhibition Therapy

Suitable for the treatment of infectious diseases, cancer and inherited disease caused by inappropriate gene activity.

The aim is to introduce a gene whose product either:

- Inhibits the expression of another gene.
- Interferes with the activity of the product of another gene.

The basis of this therapy is to eliminate the activity of a gene that encourages the growth of disease-related cells.

For example, cancer is sometimes the result of the over-activation of an oncogene (gene which stimulates cell growth). So, by eliminating the activity of that oncogene through gene inhibition therapy, it is possible to prevent further cell growth and stop the cancer in its tracks.

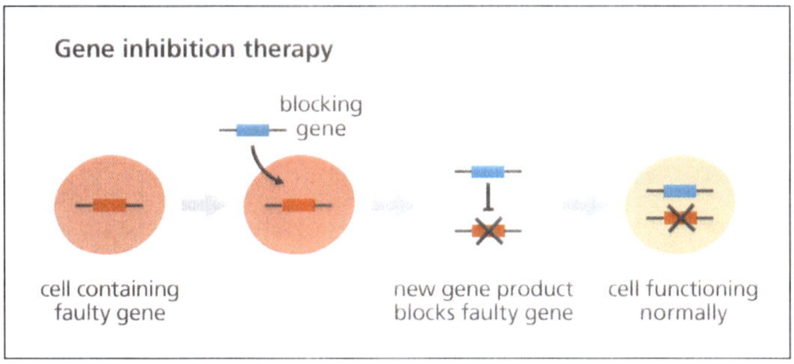

Killing of Specific Cells

Suitable for diseases such as cancer that can be treated by destroying certain groups of cells.

The aim is to insert DNA into a diseased cell that causes that cell to die.

This can be achieved in one of two ways:

- The inserted DNA contains a "suicide" gene that produces a highly toxic product which kills the diseased cell.
- The inserted DNA causes expression of a protein that marks the cells so that the diseased cells are attacked by the body's natural immune system.

It is essential with this method that the inserted DNA is targeted appropriately to avoid the death of cells that are functioning normally.

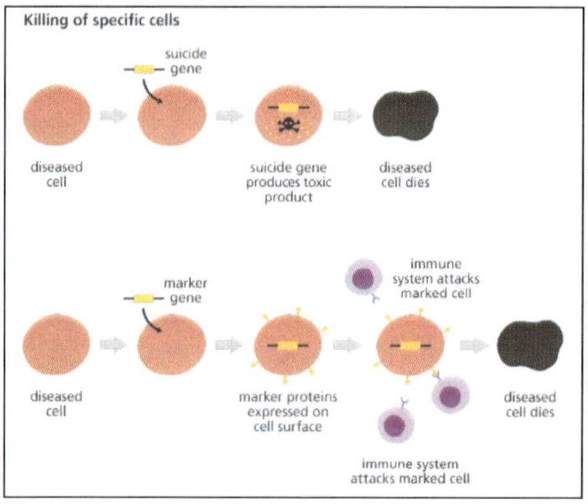

DNA Transfer

A section of DNA/gene containing instructions for making a useful protein is packaged within a vector, usually a virus, bacterium or plasmid. The vector acts as a vehicle to carry the new DNA into the cells of a patient with a genetic disease.

Once inside the cells of the patient, the DNA/gene is expressed by the cell's normal machinery leading to production of the therapeutic protein and treatment of the patient's disease.

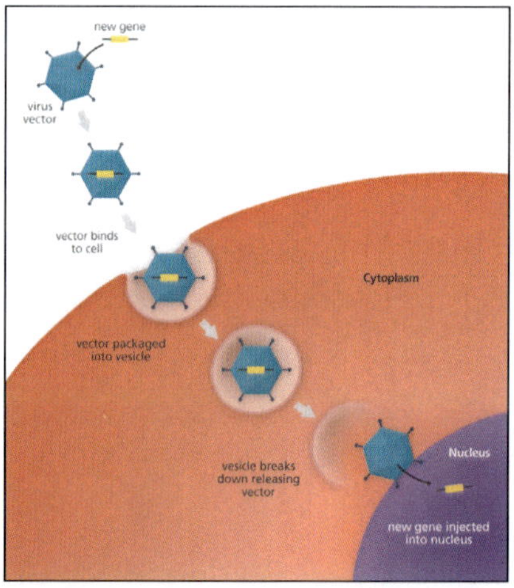

An illustration to show the transfer of a new gene into the nucleus of a cell via a viral vector.

Challenges of Gene Therapy

Delivering the gene to the right place and switching it on:

- It is crucial that the new gene reaches the right cell.
- Delivering a gene into the wrong cell would be inefficient and could also cause health problems for the patient.
- Even once the right cell has been targeted the gene has to be turned on.
- Cells sometimes obstruct this process by shutting down genes that are showing unusual activity.

Avoiding the immune response:

- The role of the immune system is to fight off intruders.
- Sometimes new genes introduced by gene therapy are considered potentially-harmful intruders.
- This can spark an immune response in the patient, that could be harmful to them.
- Scientists therefore have the challenge of finding a way to deliver genes without the immune system 'noticing'.
- This is usually by using vectors that are less likely to trigger an immune response.

Making sure the new gene doesn't disrupt the function of other genes:

- Ideally, a new gene introduced by gene therapy will integrate itself into the genome of the patient and continue working for the rest of their lives.
- There is a risk that the new gene will insert itself into the path of another gene, disrupting its activity.
- This could have damaging effects, for example, if it interferes with an important gene involved in regulating cell division, it could result in cancer.
- The cost of gene therapy:
- Many genetic disorders that can be targeted with gene therapy are extremely rare.
- Gene therapy therefore often requires an individual, case-by-case approach. This may be effective, but may also be very expensive.

Whole Genome Sequencing

Whole genome sequencing is the act of deducing the complete nucleic acid sequence of the genetic code, or genome, of an organism or organelle (specifically, the mitochondrion or chloroplast). The first whole genome sequencing efforts, carried out in 1976 and 1977, focused respectively on the bacteriophages (bacteria-infecting viruses) MS2 and ΦX174, which have relatively small genomes. Since then there have been numerous innovations in the field of DNA sequencing that have expanded the capabilities of the technology. Those innovations, combined with increasing cost-effectiveness in the early 21st century, enabled the routine use of whole genome sequencing in laboratories worldwide, which effectively ushered in a new era of biological discovery. The power of the approach has been realized in the study of human populations and human diseases such as cancer, as well as in the elucidation of whole genome sequences of crop plants, livestock, and other species of scientific or agricultural significance. Thus, it is acknowledged generally that there exists great value in a detailed understanding of the nucleic acid sequence-especially the variations in the sequence that correlate with predisposition to health or disease states or with other properties of societal or economic significance in microbial, animal, and plant populations.

Sequencing Methods: From Genes to Genomes

In 1944 Canadian-born American bacteriologist Oswald Avery and colleagues recognized that the hereditary material passed from parent to offspring was DNA. Subsequent genetic analyses carried out by other scientists on viruses, bacteria, yeast, fruit flies, and nematodes demonstrated that the intentional induction of mutations that disrupted the genetic code, combined with the analysis of observable traits (phenotypes) produced by such mutations, were important approaches to the study of gene function. Such studies, however, were able to query only a fraction of genes in a genome.

The first sequencing methods (the Maxam-Gilbert and Sanger methods), developed in the 1970s, were deployed to reveal the nucleic acid composition of individual genes and the relatively small genomes of certain viruses. The sequencing of larger genomes remained out of reach conceptually, because of high costs and the effort required, until the launch of the Human Genome Project (HGP) in 1990 in the United States. Although the project was not universally embraced, some recognized that technology had evolved to the point where whole genome sequencing of larger genomes could be considered realistically. Particularly important was the development of automated sequencing machines that employed fluorescence instead of radioactive decay for the detection of the sequencing reaction products. Automation offered new possibilities for scaling up the production of DNA sequencing to tackle large genomes.

An early aim of the HGP was to obtain the whole genome sequences of important experimental model organisms, such as the yeast Saccharomyces cerevisiae, the fruit fly

Drosophila melanogaster, and the nematode Caenorhabditis elegans. In sequencing those smaller and therefore more-tractable genomes, three outcomes were anticipated. First, the sequences would be of value to the research community, serving to accelerate efforts to understand gene function by using model systems. Second, the experience gained would inform approaches to sequencing the human genome and other similarly sized genomes. Third, functional relationships between sequences of different organisms would be revealed as a consequence of cross-species sequence similarity. Ultimately, with the involvement of more than one thousand scientists worldwide, two human genome sequences were published in 2001. With this development came established methods and analytic standards that were used to sequence other large genomes.

A major challenge for de novo sequencing, in which sequences are assembled for the very first time (such as with the HGP), is the production of individual DNA reads that are of sufficient length and quality to span common repetitive elements, which are a general property of complex genome sequences and a source of ambiguity for sequence assembly. In many of the early de novo whole genome sequencing projects, emphasis was placed on the production of so-called reference sequences, which were of enduring high quality and would serve as the foundation for future experimentation.

An important approach used by many projects that sequenced large genomes involved hierarchical shotgun sequencing, in which segments of genomic DNA were cloned (copied) and arranged into ordered arrays. Those ordered arrays were known as physical maps, and they served to break large genomes into thousands of short DNA fragments. Those short fragments were then aligned, such that identical sequences overlapped, thereby enabling the fragments to be linked together to yield the full-length genomic sequence. The fragments were relatively easy to manipulate in the laboratory, could be apportioned among collaborating laboratories, and were amenable to the detailed error-correction exercises important in generating the high-quality reference sequences sought by HGP scientists. Some genome projects were conducted without the use of such maps, using instead an approach called whole genome shotgun sequencing. This approach avoided the time and expense needed to create physical maps and provided more-rapid access to the DNA sequence.

Whether using physical maps or the whole genome shotgun sequencing approach, the sequencing exercise involved randomly fragmenting either cloned (copied) or native genomic DNA into very short segments that could then be inserted into bacterial cells as plasmids for amplification, producing many copies of the segments, prior to nucleic acid purification and sequence analysis. In a process known as assembly, computer programs were then used to stitch the sequences back together to reconstruct the original DNA sequencing target. Assembly of whole genome shotgun sequencing data was difficult and required sophisticated computer programs and powerful supercomputers, and, even in the years following the completion of the HGP, whole genome shotgun sequence assembly remained a significant challenge for whole genome sequencing projects.

Next-Generation Technologies

Although the first whole genome sequences were in themselves technological and scientific feats of significance, the scientific opportunities and the host of technologies those projects spawned have had even greater impacts. Among the most significant technological developments has been in the area of next-generation DNA sequencing technologies for human genome analysis. Certain of those technologies originally were designed to re-sequence genomes (as opposed to de novo sequencing). In re-sequencing, short sequences are produced and aligned computationally to existing reference genome sequences generated, at least initially, using the older de novo sequencing methods. Next-generation sequencing approaches are characterized generally by the massively parallel production of short sequences, in which multiple DNA fragments are generated simultaneously and in sufficient quantity to redundantly represent every base in the target genome.

Polymerase Chain Reaction

Polymerase chain reaction (PCR) is a common laboratory technique used to make many copies (millions or billions!) of a particular region of DNA. This DNA region can be anything the experimenter is interested in. For example, it might be a gene whose function a researcher wants to understand, or a genetic marker used by forensic scientists to match crime scene DNA with suspects.

Typically, the goal of PCR is to make enough of the target DNA region that it can be analyzed or used in some other way. For instance, DNA amplified by PCR may be sent for sequencing, visualized by gel electrophoresis, or cloned into a plasmid for further experiments.

PCR is used in many areas of biology and medicine, including molecular biology research, medical diagnostics, and even some branches of ecology.

Taq Polymerase

Like DNA replication in an organism, PCR requires a DNA polymerase enzyme that makes new strands of DNA, using existing strands as templates. The DNA polymerase typically used in PCR is called Taq polymerase, after the heat-tolerant bacterium from which it was isolated (Thermus aquaticus).

T. aquaticus lives in hot springs and hydrothermal vents. Its DNA polymerase is very heat-stable and is most active around 70 °C (a temperature at which a human or E. coli DNA polymerase would be nonfunctional). This heat-stability makes Taq polymerase ideal for PCR. high temperature is used repeatedly in PCR to denature the template DNA, or separate its strands.

PCR Primers

Like other DNA polymerases, Taq polymerase can only make DNA if it's given a primer, a short sequence of nucleotides that provides a starting point for DNA synthesis. In a PCR reaction, the experimenter determines the region of DNA that will be copied, or amplified, by the primers she or he chooses.

PCR primers are short pieces of single-stranded DNA, usually around 202020 nucleotides in length. Two primers are used in each PCR reaction, and they are designed so that they flank the target region. That is, they are given sequences that will make them bind to opposite strands of the template DNA, just at the edges of the region to be copied. The primers bind to the template by complementary base pairing.

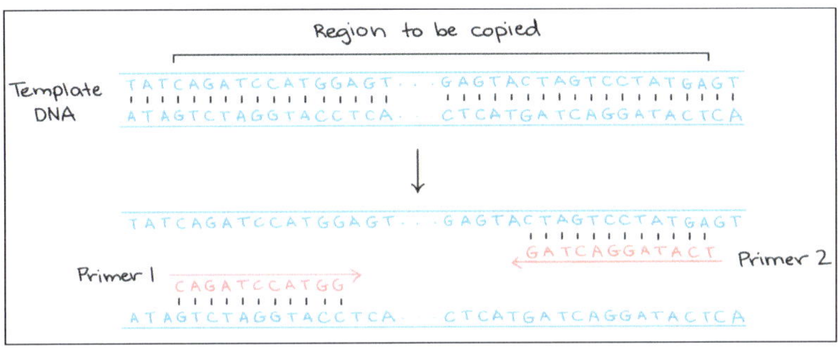

When the primers are bound to the template, they can be extended by the polymerase, and the region that lies between them will get copied.

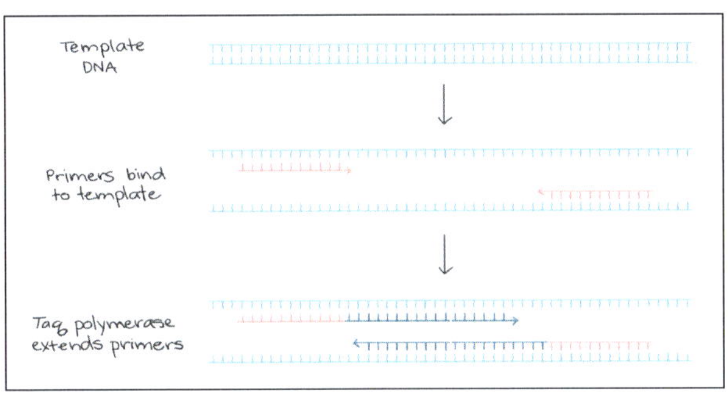

Steps of PCR

The key ingredients of a PCR reaction are Taq polymerase, primers, template DNA, and nucleotides (DNA building blocks). The ingredients are assembled in a tube, along with cofactors needed by the enzyme, and are put through repeated cycles of heating and cooling that allow DNA to be synthesized.

The basic steps are:

1. Denaturation (96 °C): Heat the reaction strongly to separate, or denature, the DNA strands. This provides single-stranded template for the next step.

2. Annealing (55 – 65 °C): Cool the reaction so the primers can bind to their complementary sequences on the single-stranded template DNA.

3. Extension (72 °C): Raise the reaction temperatures so Taq polymerase extends the primers, synthesizing new strands of DNA.

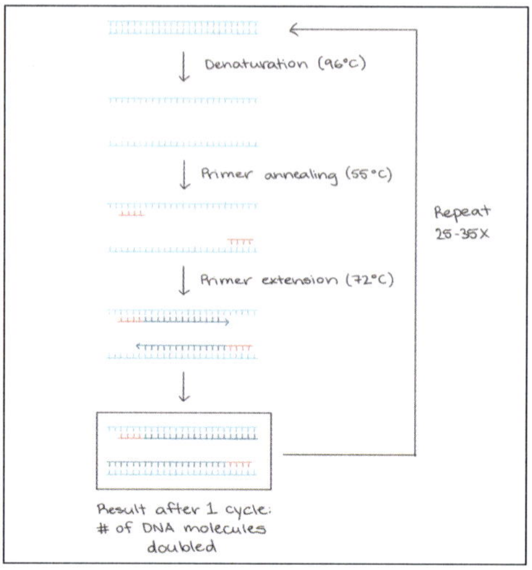

This cycle repeats 25 - 35 times in a typical PCR reaction, which generally takes 2 - 4 hours, depending on the length of the DNA region being copied. If the reaction is efficient (works well), the target region can go from just one or a few copies to billions.

That's because it's not just the original DNA that's used as a template each time. Instead, the new DNA that's made in one round can serve as a template in the next round of DNA synthesis. There are many copies of the primers and many molecules of *Taq* polymerase floating around in the reaction, so the number of DNA molecules can roughly double in each round of cycling. This pattern of exponential growth is shown in the image below.

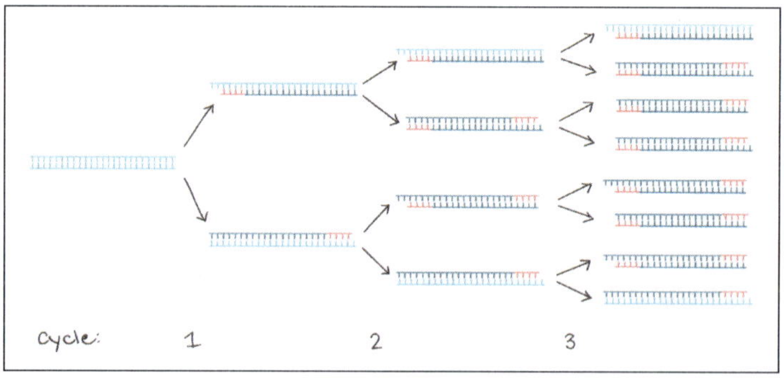

Using Gel Electrophoresis to Visualize the Results of PCR

The results of a PCR reaction are usually visualized (made visible) using gel electrophoresis. Gel electrophoresis is a technique in which fragments of DNA are pulled through a gel matrix by an electric current, and it separates DNA fragments according to size. A standard, or DNA ladder, is typically included so that the size of the fragments in the PCR sample can be determined.

DNA fragments of the same length form a "band" on the gel, which can be seen by eye if the gel is stained with a DNA-binding dye. For example, a PCR reaction producing a 400 base pair (bp) fragment would look like this on a gel:

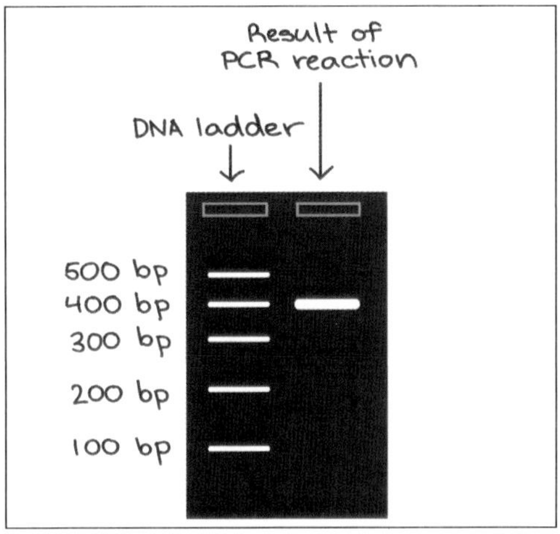

A DNA band contains many, many copies of the target DNA region, not just one or a few copies. Because DNA is microscopic, lots of copies of it must be present before we can see it by eye. This is a big part of why PCR is an important tool: it produces enough copies of a DNA sequence that we can see or manipulate that region of DNA.

Applications of PCR

Using PCR, a DNA sequence can be amplified millions or billions of times, producing enough DNA copies to be analyzed using other techniques. For instance, the DNA may be visualized by gel electrophoresis, sent for sequencing, or digested with restriction enzymes and cloned into a plasmid.

PCR is used in many research labs, and it also has practical applications in forensics, genetic testing, and diagnostics. For instance, PCR is used to amplify genes associated with genetic disorders from the DNA of patients (or from fetal DNA, in the case of prenatal testing). PCR can also be used to test for a bacterium or DNA virus in a patient's body: if the pathogen is present, it may be possible to amplify regions of its DNA from a blood or tissue sample.

References

- The new encyclopaedia Britannica. Encyclopaedia Britannica, inc. (15th ed.). Chicago: Encyclopaedia Britannica. 1993. p. 178. ISBN 0852295715. OCLC 27665641
- What-is-gmo: nongmoproject.org, Retrieved 20 May, 2019

- Panesar, Pamit et al. (2010) Enzymes in Food Processing: Fundamentals and Potential Applications, Chapter 10, I K International Publishing House, ISBN 978-93-80026-33-6
- What-is-genetically-modified-food: vox.com, Retrieved 27 January, 2019
- Campbell-Platt, Geoffrey (26 August 2011). Food Science and Technology. Ames, Iowa: John Wiley & Sons. ISBN 978-1-4443-5782-0.
- What-is-gene-therapy: yourgenome.org, Retrieved 7 March, 2019
- Bull JJ, Smithson MW, Nuismer SL (January 2018). "Transmissible Viral Vaccines". Trends in Microbiology. 26 (1): 6–15. doi:10.1016/j.tim.2017.09.007. PMC 5777272. PMID 29033339
- Whole-genome-sequencing, science: britannica.com, Retrieved 30 August, 2019
- Duncan, R. (1996). "Tissue Culture-Induced Variation and Crop Improvement". Advances in Agronomy Volume 58. Advances in Agronomy. 58. pp. 201–40. doi:10.1016/S0065-2113(08)60256-4. ISBN 9780120007585
- Polymerase-chain-reaction-pcr, dna-sequencing-pcr-electrophoresis, biotech-dna-technology, biology, science: khanacademy.org, Retrieved 25 February, 2019

Permissions

All chapters in this book are published with permission under the Creative Commons Attribution Share Alike License or equivalent. Every chapter published in this book has been scrutinized by our experts. Their significance has been extensively debated. The topics covered herein carry significant information for a comprehensive understanding. They may even be implemented as practical applications or may be referred to as a beginning point for further studies.

We would like to thank the editorial team for lending their expertise to make the book truly unique. They have played a crucial role in the development of this book. Without their invaluable contributions this book wouldn't have been possible. They have made vital efforts to compile up to date information on the varied aspects of this subject to make this book a valuable addition to the collection of many professionals and students.

This book was conceptualized with the vision of imparting up-to-date and integrated information in this field. To ensure the same, a matchless editorial board was set up. Every individual on the board went through rigorous rounds of assessment to prove their worth. After which they invested a large part of their time researching and compiling the most relevant data for our readers.

The editorial board has been involved in producing this book since its inception. They have spent rigorous hours researching and exploring the diverse topics which have resulted in the successful publishing of this book. They have passed on their knowledge of decades through this book. To expedite this challenging task, the publisher supported the team at every step. A small team of assistant editors was also appointed to further simplify the editing procedure and attain best results for the readers.

Apart from the editorial board, the designing team has also invested a significant amount of their time in understanding the subject and creating the most relevant covers. They scrutinized every image to scout for the most suitable representation of the subject and create an appropriate cover for the book.

The publishing team has been an ardent support to the editorial, designing and production team. Their endless efforts to recruit the best for this project, has resulted in the accomplishment of this book. They are a veteran in the field of academics and their pool of knowledge is as vast as their experience in printing. Their expertise and guidance has proved useful at every step. Their uncompromising quality standards have made this book an exceptional effort. Their encouragement from time to time has been an inspiration for everyone.

The publisher and the editorial board hope that this book will prove to be a valuable piece of knowledge for students, practitioners and scholars across the globe.

Index